Life in the Universe

Life in the Universe

. . .

SCIENTIFIC AMERICAN : A SPECIAL ISSUE

W. H. FREEMAN AND COMPANY
New York

Library of Congress Cataloging–in–Publication Data

Life in the universe : readings from Scientific American magazines
p. cm.
Includes bibliographical references and index.
ISBN 0-7167-2714-5 (pbk.)
1. Life on other planets. 2. Life—Origin. 3. Earth—Origin. 4. Scientific American
I. Scientific American.
QB54.L4845 1995
577—dc20 94-44733
CIP

The ten chapters and the epilogue in this book originally appeared as articles and the closing essay in the October 1994 issue of SCIENTIFIC AMERICAN.

Printed in the United States of America

1 2 3 4 5 6 7 8 9 0 RRD 9 9 8 7 6 5 4 3 2

CONTENTS

Life in the Universe

Life in the Universe

We comprehend the universe and our place in it. But there are limits to what we can explain at present. Will research at the boundaries of science reveal a special role for intelligent life?

• • •

Steven Weinberg

In Walt Whitman's often quoted poem "When I Heard the Learn'd Astronomer," the poet tells how, being shown the astronomer's charts and diagrams, he became tired and sick and wandered off by himself to look up "in perfect silence at the stars." Generations of scientists have been annoyed by these lines. The sense of beauty and wonder does not become atrophied through the work of science, as Whitman implies. The night sky is as beautiful as ever, to astronomers as well as to poets. And as we understand more and more about nature, the scientist's sense of wonder has not diminished but has rather become sharper, more narrowly focused on the mysteries that still remain.

The nearby stars that Whitman could see without a telescope are now not so mysterious. Massive computer codes simulate the nuclear reactions at the stars' cores and follow the flow of energy by convection and radiation to their visible surfaces, explaining both their present appearance and how they have evolved. The observation in 1987 of gamma rays and neutrinos from the supernova in the Large Magellanic Cloud provided dramatic confirmation of the theory of stellar structure and evolution. These theories are themselves beautiful to us, and knowing why Betelgeuse is red may even add to the pleasure of looking at the winter sky.

But there are plenty of mysteries left, many of them discussed by other authors in this book. Of what kind of matter are galaxies and galactic clusters made? How did the stars, planets and galaxies form? How widespread in the universe are habitats suitable for life? How did the earth's oceans and atmosphere form? How did life start? What are the relations of cause and effect between the evolution of life and the terrestrial environment in which it has occurred? How large is the role of chance in the origin of the human species? How does the brain think? How do human institutions respond to environmental and technological change?

We may be very far from the solution of some of these problems. Still, we can guess what kinds of solutions they will have, in a way that was not possible when *Scientific American* was founded 150 years ago. New ideas and insights will be needed, which we can expect to find within the boundaries of science as we know it.

Then there are mysteries at the outer boundaries of our science, matters that we cannot hope to explain in terms of what we already know. When we explain anything we observe, it is in terms of scientific principles that are themselves explained in terms of deeper principles. Following this chain of explanations, we are led at last to laws of nature that cannot be explained within the boundaries of contemporary science. And in dealing with life and many other aspects of nature, our explanations have a historical component. Some historical facts are accidents that can never be explained, except perhaps statistically: we can never explain precisely why life on the earth takes the form it does, although we can hope to show that some forms are more likely than others. We can explain a great deal, even where history plays a role, in terms of the conditions with which the universe began, as well as the laws of nature. But how do we explain the initial conditions? A further complex of puzzles overhangs the laws of nature and the initial conditions. It concerns the dual role of intelligent life—as part of the universe we seek to explain, and as the explainer.

The laws of nature as we currently understand them allow us to trace the observed expansion of the universe back to what would be a true beginning, a moment when the universe was infinitely hot and dense, some 10 to 20 billion years ago. We do not have enough confidence in the applicability of these laws at extreme temperatures and densities to be sure that there really was such a moment, much less to work out all the initial conditions, if there were any. For the present, we cannot do better than to describe the initial conditions of the universe at a time about 10^{-12} second after the nominal moment of infinite temperature (see figure "A Timeline for the History of Life in the Universe").

The temperature of the universe had dropped by then to about 10^{15} degrees, cool enough for us to apply our physical theories. At these temperatures the universe would have been filled with a gas consisting of all the types of particles known to high-energy nuclear physics, together with their antiparticles, continually being annihilated and cre-

A Timeline for the History of Life in the Universe

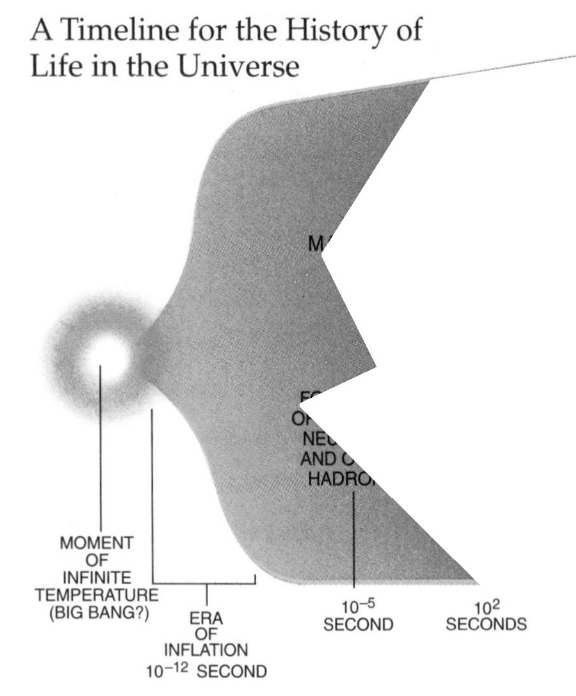

ated in their collisions. As the universe continued to expand and cool, creation became slower than annihilation, and almost all the particles and antiparticles disappeared. If there had not been a small excess of electrons over antielectrons, and quarks over antiquarks, then ordinary particles like electrons and quarks would be virtually absent in the universe today. It is this early excess of matter over antimatter, estimated as one part in about 10^{10}, that survived to form light atomic nuclei three minutes later, then after a million years to form atoms and later to be cooked to heavier elements in stars, ultimately to provide the material out of which life would arise. The one part in 10^{10} excess of matter over antimatter is one of the key initial conditions that determined the future development of the universe.

In addition, there may exist other types of particles, not yet observed in our laboratories, that interact more weakly with one another than do quarks and electrons and that therefore would have annihilated relatively slowly. Large numbers of these exotic particles would have been left over from the early universe, forming the "dark matter" that now apparently makes up much of the mass of the universe.

Finally, although it is generally assumed that when the universe was 10^{-12} second old its contents were pretty nearly the same everywhere, small inhomogeneities must have existed that triggered the formation, millions of years later, of the first galaxies and stars. We cannot directly observe any inhomogeneities at times earlier than about a million years after the beginning, when the universe first became transparent. Astronomers are currently engaged in mapping minute variations in the intensity of the cosmic microwave radiation background that was emitted at that time, using them to infer the primordial distribution of matter.

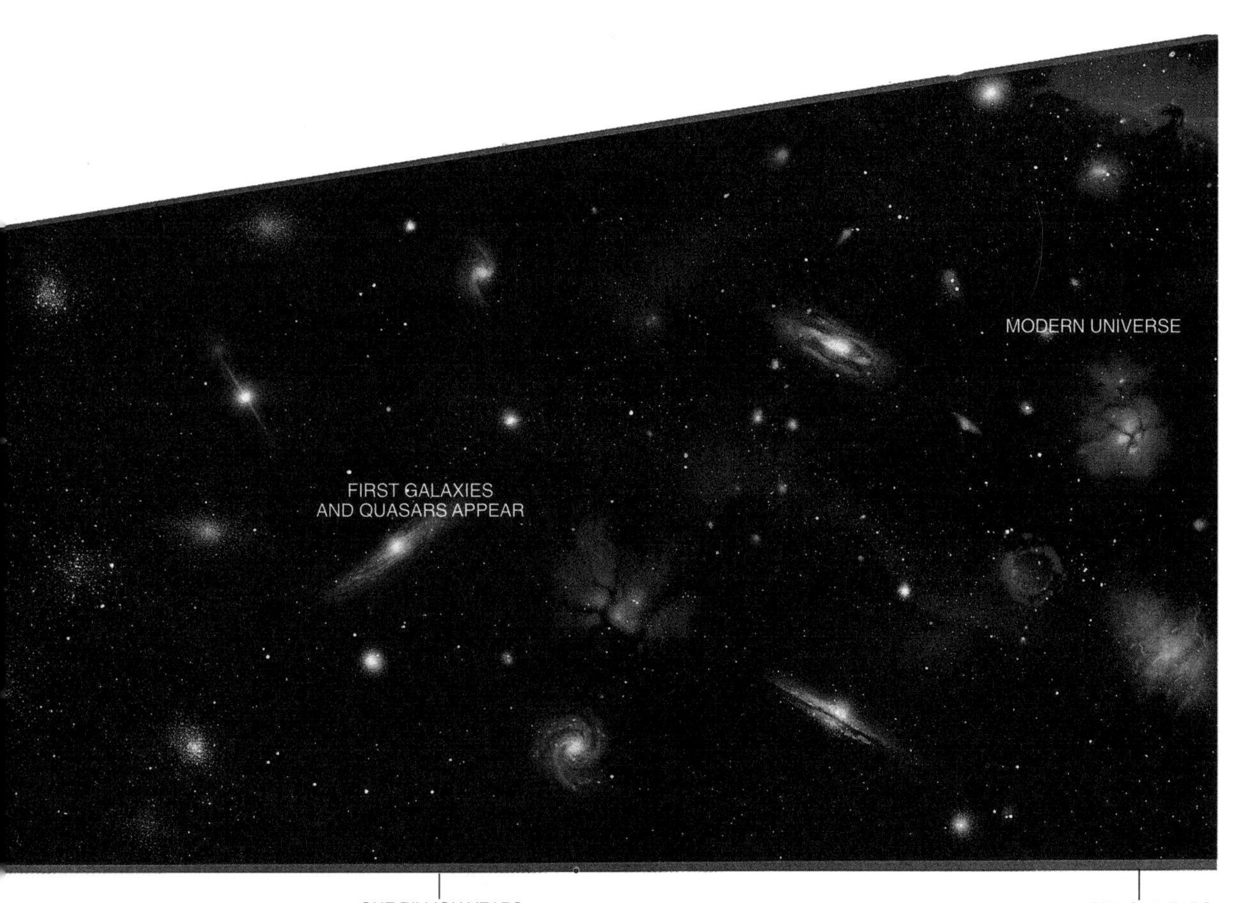

This information can in turn be used to deduce the initial inhomogeneities at 10^{-12} second after the beginning.

From the austere viewpoint of fundamental physics, the history of the universe is just an illustrative example of the laws of nature. At the deepest level to which we have been able to trace our explanations, those laws take the form of quantum field theories. When quantum mechanics is applied to a field such as the electromagnetic field, it is found that the energy and momentum of the field come in bundles, or quanta, that are observed in the laboratory as particles. The modern Standard Model posits an electromagnetic field, whose quanta are photons; an electron field, whose quanta are electrons and antielectrons; and a number of other fields whose quanta are particles called leptons and antileptons. There are various quark fields whose quanta are quarks and antiquarks, and there are 11 other fields whose quanta are the particles that transmit the weak and strong forces that act on the elementary particles.

The Standard Model is certainly not the final law of nature. Even in its simplest form it contains a number of arbitrary features. Some 18 numerical parameters exist whose values have to be taken from experiment, and the multiplicity of types of quarks and leptons is unexplained. Also, one aspect of the model is still uncertain: we are not sure of the details of the mechanism that gives masses to the quarks, electrons and other particles. This is the puzzle that was to have been solved by the now canceled Superconducting Super Collider. We hope it will be unraveled by the Large Hadron Collider being planned at CERN near Geneva. Finally, the model is incomplete; it does not include gravitation. We have a good field theory of gravitation, the General Theory of Relativity, but the quantum version of this theory breaks down at very high energies.

It is possible that all these problems will find their solution in a new kind of theory known as string theory. The point particles of quantum field theory are reinterpreted in string theory as tiny, extended one-dimensional objects called strings. These strings can exist in various modes of vibration, each mode appearing in the laboratory as a different type of particle. String theory not only provides a quantum description of gravitation that makes sense at all energies; one of the modes of vibration of a string would appear as a particle with the properties of the graviton, the quantum of the gravitational field, so string theory even offers an explanation of why gravitation exists. Further, there are versions of string theory that predict something like the menu of fields incorporated in the Standard Model.

But string theory has had no successes yet in explaining or predicting any of the numerical parameters of the Standard Model. Moreover, strings are much too small for us to detect directly the stringy nature of elementary particles; a string is smaller relative to an atomic nucleus than is a nucleus relative to a mountain. The intellectual investment now being made in string theory without the slightest encouragement from experiment is unprecedented in the history of science. Yet for now, it offers our best hope for a deeper understanding of the laws of nature.

The present gaps in our knowledge of the laws of nature stand in the way of explaining the initial conditions of the universe, at 10^{-12} second after the nominal beginning, in terms of the history of the universe at earlier times. Calculations in the past few years have made it seem likely that the tiny excess of quarks and electrons over antiquarks and antielectrons at this time was produced a little earlier, at a temperature of about 10^{16} degrees. At that moment the universe went through a phase transition, something like the freezing of water, in which the known elementary particles for the first time acquired mass. But we cannot explain why the excess produced in this way should be one part in 10^{10}, or calculate its precise value, until we understand the details of the mass-producing mechanism.

The other initial condition, the degree of inhomogeneity in the early universe, may trace back to even earlier times. In our quantum field theories of elementary particles, including the simplest version of the Standard Model, several fields pervade the universe, taking nonzero values even in supposedly empty space. In the present state of the universe, these fields have reached equilibrium values, which minimize the energy density of the vacuum. This vacuum energy density, also known as the cosmological constant, can be measured through the gravitational field that it produces. It is apparently very small.

In some modern theories of the early universe, however, there was a very early time when these fields had not yet reached their equilibrium values, so that the vacuum would have had an enormous energy density. This energy would have produced a rapid expansion of the universe, known as inflation. Tiny inhomogeneities that would have been produced by quantum fluctuations before this inflation would have been magnified in the expansion and could have produced the much larger inhomogeneities that millions of years later triggered the formation of galaxies. It has even been conjectured that the inflation that began the expansion of the visible universe did not occur throughout the cosmos. It may instead have been just one local episode in an eternal succession of local inflations that occur at random throughout an infinite universe. If this is true, then the problem of initial conditions disappears; there was no initial moment.

In this picture, our local expansion may have begun with some special ingredients or inhomogeneities, but like the forms of life on the earth, these could be understood only in a statistical sense. Unfortunately, at the time of inflation gravitation was so strong that quantum gravitational effects were important. So these ideas will remain speculative until we understand the quantum theory of gravitation—perhaps in terms of something like a string theory.

The experience of the past 150 years has shown that life is subject to the same laws of nature as is inanimate matter. Nor is there any evidence of a grand design in the origin or evolution of life (see figure "The Emergence of Life"). There are well-known problems in the description of consciousness in terms of the working of the brain. They arise because we each have special knowledge of our own consciousness that does not come to us from the senses. In principle, no obstacle stands in the way of explaining the *behavior* of other people in terms of neurology and physiology and, ultimately, in terms of physics and history. When we have succeeded in this endeavor, we should find that part of the explanation is a program of neural activity that we will recognize as corresponding to our own consciousness.

But as much as we would like to take a unified view of nature, we keep encountering a stubborn duality in the role of intelligent life in the universe (see figure "The Emergence of Intelligence"), as both subject and student. We see this even at the deepest level of modern physics. In quantum mechanics the state of any system is described by a mathematical object known as the wave function. According to the interpretation of quantum mechanics worked out in Copenhagen in the early 1930s, the rules for calculating the wave function are of a very different character from the principles used to interpret it. On one hand, there is the Schrödinger equation, which describes in a perfectly deterministic way how the wave function of any system changes with time. Then, quite separate, there is a set of principles that tells how to use the wave function to calculate the probabilities of various possible outcomes when someone makes a measurement.

The Copenhagen interpretation holds that when we measure any quantity, such as position or momentum, we are intervening in a way that causes an unpredictable change in the wave function, resulting in a wave function for which the measured quantity has some definite value, in a manner that cannot be described by the deterministic Schrödinger equation. For instance, before a measurement the wave function of a spinning electron is generally a sum of terms corresponding to different directions of the electron's spin; in such a state the electron cannot be said to be spinning in any particular direction. If we measure whether the electron is spinning clockwise or counterclockwise around some axis, however, we somehow change the electron's wave function so that it is definitely spinning one way or the other. Measurement is thus regarded as something intrinsically different from anything else in nature. And although opinions differ, it is hard to identify anything special that qualifies some process to be called a measurement, except its effect on a conscious mind.

Among physicists and philosophers one finds at least four different reactions to the Copenhagen interpretation. The first is simply to accept it as it stands. This attitude is mostly limited to those who are attracted to the old, dualistic worldview that puts life and consciousness on a different footing from the rest of nature. The second attitude is to accept the rules of the Copenhagen interpretation for practical purposes, without worrying about their ultimate interpretation. This attitude is by far the most common among working physicists. The third approach is to try to avoid these problems by changing quantum mechanics in some way. So far no such attempt has found much acceptance among physicists.

The final approach is to take the Schrödinger equation seriously, to give up the dualism of the Copenhagen interpretation and to try to explain its successful rules through a description of measurers and their apparatus in terms of the same deterministic evolution of the wave function that governs everything else. When we measure some quantity (like the direction of an electron's spin), we put the system in an environment (for instance, a magnetic field) where its energy (or momentum) has a strong dependence on the value of the measured quantity. According to the Schrödinger equation, the different terms in the wave function that correspond to different energies will oscillate at rates proportional to these energies.

A measurement thus makes the terms of the wave function that correspond to different values of a measured quantity, such as an electron spin, oscillate rapidly at different rates, so they cannot interfere with one another in any future measurement, just as the signals from radio stations broadcasting at widely spaced frequencies do not interfere. In this way, a measurement causes the history of the universe for practical purposes to diverge into different noninterfering tracks, one for each possible value of the measured quantity.

Yet how do we explain the Copenhagen rules for calculating the probabilities for these different "worldtracks" in a world governed by the completely deterministic Schrödinger equation? Progress has recently been made on this problem, but it is not yet definitely solved. (For what it is worth, I prefer this last approach, although the second has much to recommend it.)

It is also difficult to avoid talking about living observers when we ask why our physical principles are what they are. Modern quantum field theory and string theory can be understood as answers to the problem of reconciling quantum mechanics and special relativity in such a way that experiments are guaranteed to give sensible results. We require that the results of our dynamical calculations must satisfy conditions known to field theorists as unitarity, positivity and cluster decomposition. Roughly speaking, these conditions require that probabilities always add up to 100 percent, that they are always positive and that those observed in distant experiments are not related.

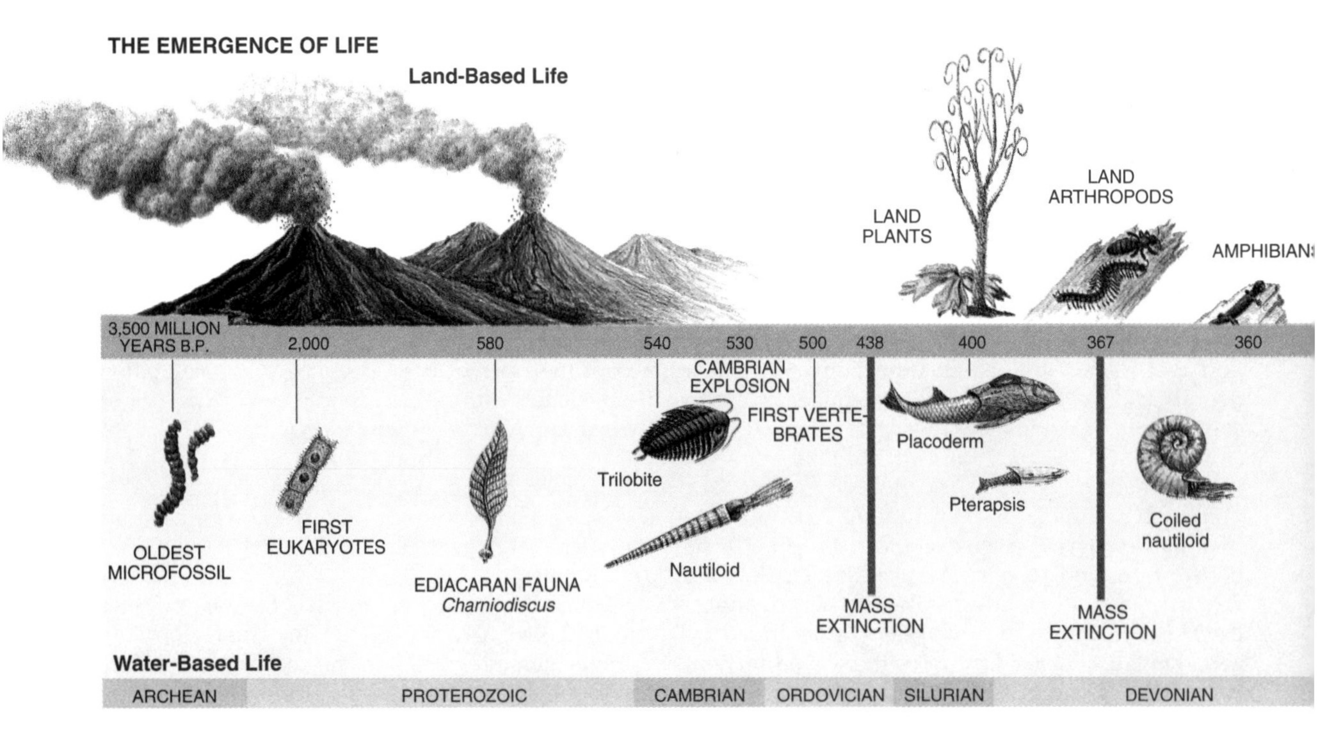

This is not so easy. If we try to write down some dynamical equations that will automatically give results consistent with some of these conditions, we usually find that the results violate the other conditions. It seems that any relativistic quantum theory that satisfies all these conditions must appear at sufficiently low energy like a quantum field theory. That is presumably why nature at accessible energies is so well described by the quantum field theory known as the Standard Model.

Also, so far as we can tell, the only mathematically consistent relativistic quantum theories that satisfy these conditions at all energies and that involve gravitation are string theories. Further, the student of string theory who asks why one makes this or that mathematical assumption is told that otherwise one would violate physical principles like unitarity and positivity. But why are these the correct conditions to impose on the results of all imaginable experiments if the laws of nature allow the possibility of a universe that contains no living beings to carry out experiments?

This question does not intrude on much of the actual work of theoretical physics, but it becomes urgent when we seek to apply quantum mechanics to the whole universe. At present, we do not understand even in principle how to calculate or interpret the wave function of the universe, and we cannot resolve these problems by requiring that all experiments should give sensible results, because by definition there is no observer outside the universe who can experiment on it.

These mysteries are heightened when we reflect how surprising it is that the laws of nature and the initial conditions of the universe should allow for the existence of beings who could observe it. Life as we know it would be impossible if any one of several physical quantities had slightly different values. The best known of these quantities is the energy of one of the excited states of the carbon 12 nucleus. There is an essential step in the chain of nuclear reactions that build up heavy elements in stars. In this step, two helium nuclei join

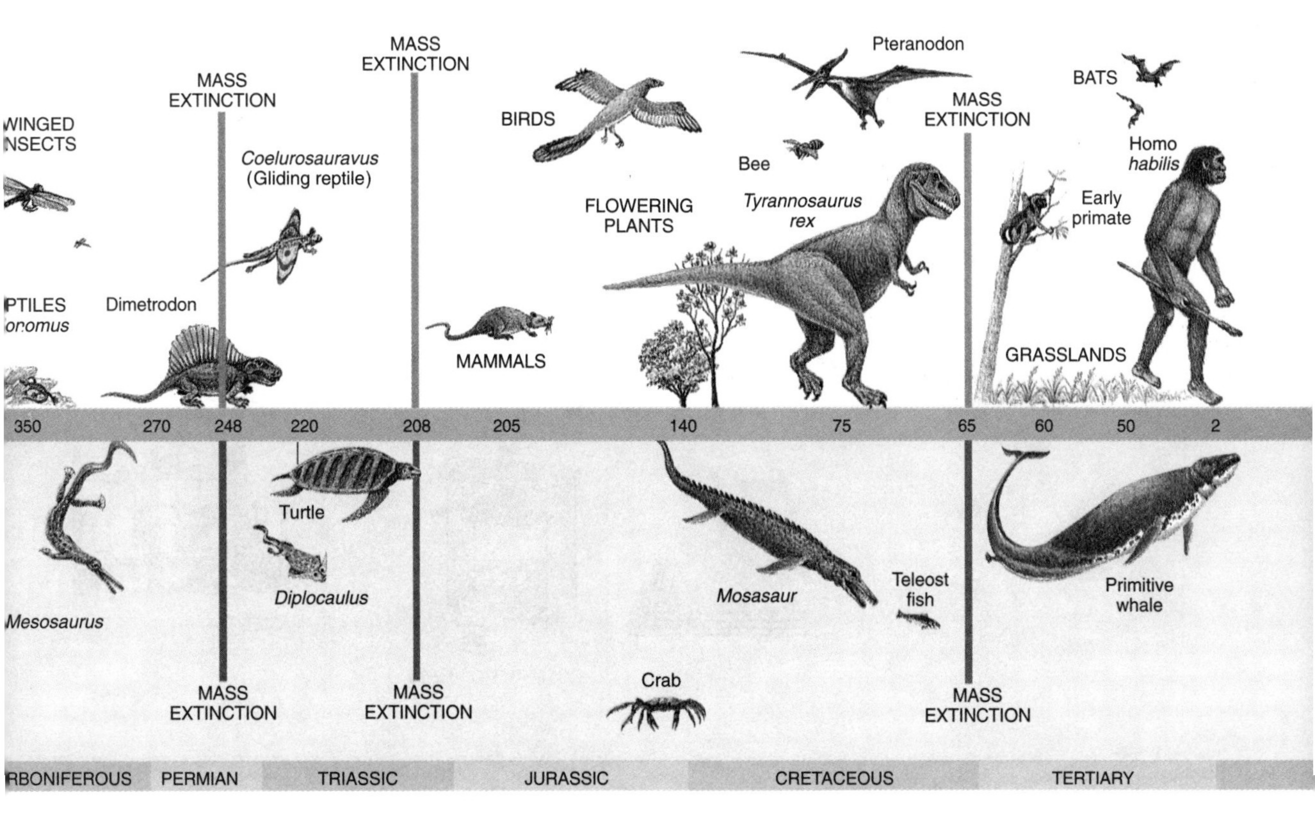

together to form the unstable nucleus of beryllium 8, which sometimes before fissioning absorbs another helium nucleus, forming carbon 12 in this excited state. The carbon 12 nucleus then emits a photon and decays into the stable state of lowest energy. In subsequent nuclear reactions carbon is built up into oxygen and nitrogen and the other heavy elements necessary for life. But the capture of helium by beryllium 8 is a resonant process, whose reaction rate is a sharply peaked function of the energies of the nuclei involved. If the energy of the excited state of carbon 12 were just a little higher, the rate of its formation would be much less, so that almost all the beryllium 8 nuclei would fission into helium nuclei before carbon could be formed. The universe would then consist almost entirely of hydrogen and helium, without the ingredients for life.

Opinions differ as to the degree to which the constants of nature must be fine-tuned to make life necessary. There are independent reasons to expect an excited state of carbon 12 near the resonant energy. But one constant does seem to require an incredible fine-tuning: it is the vacuum energy, or cosmological constant, mentioned in connection with inflationary cosmologies.

Although we cannot calculate this quantity, we can calculate some contributions to it (such as the energy of quantum fluctuations in the gravitational field that have wavelengths no shorter than about 10^{-33} centimeter). These contributions come out about 120 orders of magnitude larger than the maximum value allowed by our observations of the present rate of cosmic expansion. If the various contributions to the vacuum energy did not nearly cancel, then, depending on the value of the total vacuum energy, the universe either would go through a complete cycle of expansion and contraction before life could arise or would expand so rapidly that no galaxies or stars could form.

Thus, the existence of life of any kind seems to require a cancellation between different contributions to the vacuum energy, accurate to about 120 decimal places. It is possible that this cancellation will be explained in terms of some future theory. So far, in string theory as well as in quantum field theory, the vacuum energy involves arbitrary constants, which must be carefully adjusted to make the total vacuum energy small enough for life to be possible.

All these problems can be solved without supposing that life or consciousness plays any special

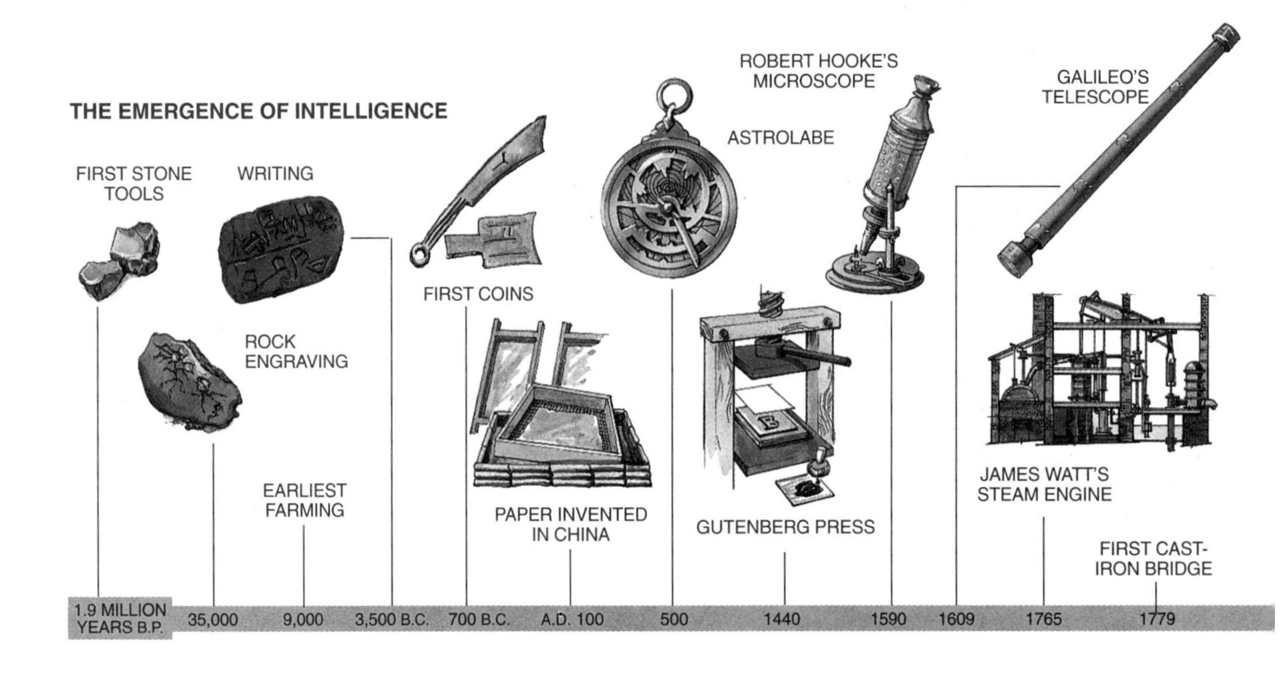

role in the fundamental laws of nature or initial conditions. It may be that what we now call the constants of nature actually vary from one part of the universe to another. (Here "different parts of the universe" could be understood in various senses. The phrase could, for example, refer to different local expansions arising from episodes of inflation in which the fields pervading the universe took different values or else to the different quantum-mechanical worldtracks that arise in some versions of quantum cosmology.) If this is the case, then it would not be surprising to find that life is possible in some parts of the universe, though perhaps not in most. Naturally, any living beings who evolve to the point where they can measure the constants of nature will always find that these constants have values that allow life to exist. The constants have other values in other parts of the universe, but there is no one there to measure them. (This is one version of what is sometimes called the anthropic principle.) Still, this presumption would not indicate any special role for life in the fundamental laws, any more than the fact that the sun has a planet on which life is possible indicates that life played a role in the origin of the solar system. The fundamental laws would be those that describe the *distribution* of values of the constants of nature between different parts of the universe, and in these laws life would play no special role.

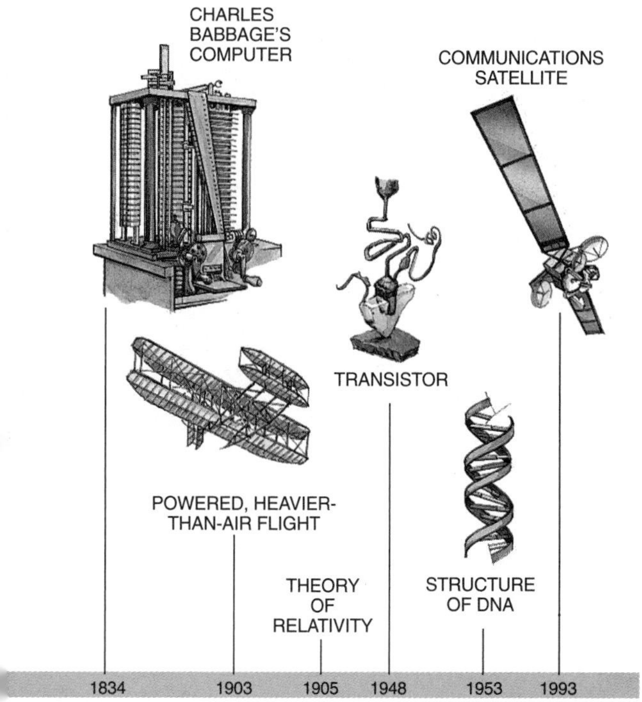

If the content of science is ultimately impersonal, its conduct is part of human culture, and not the least interesting part. Some philosophers and sociologists have gone so far as to claim that scientific principles are, in whole or in part, social constructions, like the rules of contract law or contract bridge. Most working scientists find this "social constructivist" point of view inconsistent with their own experience. Still, there is no doubt that the social context of science has become increasingly important to scientists, as we need to ask society to provide us with more and more expensive tools: accelerators, space vehicles, neutron sources, genome projects and so on.

It does not help that some politicians and journalists assume the public is interested only in those aspects of science that promise immediate practical benefits to technology or medicine. Some work on the most interesting problems of biological or physical science does have obvious practical value, but some does not, especially research that addresses problems lying at the boundaries of scientific knowledge. To earn society's support, we have to make true what we often claim: that today's basic scientific research is part of the culture of our times.

Whatever barriers now exist to communication between scientists and the public, they are not impermeable. Isaac Newton's *Principia* could at first be understood only by a handful of Europeans. Then the news that we and our universe are governed by precise, knowable laws did eventually diffuse throughout the civilized world. The theory of evolution was strenuously opposed at first; now creationists are an increasingly isolated minority. Today's research at the boundaries of science explores environments of energy and time and distance far removed from those of everyday life and often can be described only in esoteric mathematical language. But in the long run, what we learn about why the world is the way it is will become part of everyone's intellectual heritage.

The Evolution of the Universe

Some 15 billion years ago the universe emerged from a hot, dense sea of matter and energy. As the cosmos expanded and cooled, it spawned galaxies, stars, planets and life.

• • •

P. James E. Peebles, David N. Schramm, Edwin L. Turner and Richard G. Kron

At a particular instant roughly 15 billion years ago, all the matter and energy we can observe, concentrated in a region smaller than a dime, began to expand and cool at an incredibly rapid rate. By the time the temperature had dropped to 100 million times that of the sun's core, the forces of nature assumed their present properties, and the elementary particles known as quarks roamed freely in a sea of energy. When the universe had expanded an additional 1,000 times, all the matter we can measure filled a region the size of the solar system.

At that time, the free quarks became confined in neutrons and protons. After the universe had grown by another factor of 1,000, protons and neutrons combined to form atomic nuclei, including most of the helium and deuterium present today. All of this occurred within the first minute of the expansion. Conditions were still too hot, however, for atomic nuclei to capture electrons. Neutral atoms appeared in abundance only after the expansion had continued for 300,000 years and the universe was 1,000 times smaller than it is now. The neutral atoms then began to coalesce into gas clouds, which later evolved into stars. By the time the universe had expanded to one fifth its present size, the stars had formed groups recognizable as young galaxies.

When the universe was half its present size, nuclear reactions in stars had produced most of the heavy elements from which terrestrial planets were made. Our solar system is relatively young: it formed five billion years ago, when the universe was two thirds its present size. Over time the formation of stars has consumed the supply of gas in galaxies, and hence the population of stars is waning. Fifteen billion years from now stars like our sun will be relatively rare, making the universe a far less hospitable place for observers like us.

Our understanding of the genesis and evolution of the universe is one of the great achievements of 20th-century science. This knowledge comes from decades of innovative experiments and theories. Modern telescopes on the ground and in space detect the light from galaxies billions of light-years away, showing us what the universe looked like when it was young (see Figure 2.1). Particle accelerators probe the basic physics of the high-energy

Figure 2.1 GALAXY CLUSTER is representative of what the universe looked like when it was 60 percent of its present age. *The Hubble Space Telescope* captured the image by focusing on the cluster as it completed 10 orbits. This image is one of the longest and clearest exposures ever produced. Several pairs of galaxies appear to be caught in one another's gravitational field. Such interactions are rarely found in nearby clusters and are evidence that the universe is evolving.

environment of the early universe. Satellites detect the cosmic background radiation left over from the early stages of expansion, providing an image of the universe on the largest scales we can observe.

Our best efforts to explain this wealth of data are embodied in a theory known as the standard cosmological model or the big bang cosmology. The major claim of the theory is that in the large-scale average the universe is expanding in a nearly homogeneous way from a dense early state. At present, there are no fundamental challenges to the big bang theory, although there are certainly unresolved issues within the theory itself. Astronomers are not sure, for example, how the galaxies were formed, but there is no reason to think the process did not occur within the framework of the big bang. Indeed, the predictions of the theory have survived all tests to date.

Yet the big bang model goes only so far, and many fundamental mysteries remain. What was the universe like before it was expanding? (No observation we have made allows us to look back beyond the moment at which the expansion began.) What will happen in the distant future, when the last of the stars exhaust the supply of nuclear fuel? No one knows the answers yet.

Our universe may be viewed in many lights—by mystics, theologians, philosophers or scientists. In science we adopt the plodding route: we accept only what is tested by experiment or observation. Albert Einstein gave us the now well-tested and accepted Theory of General Relativity, which establishes the relations between mass, energy, space and time. Einstein showed that a homogeneous distribution of matter in space fits nicely with his theory. He assumed without discussion that the universe is static, unchanging in the large-scale average [see "How Cosmology Became a Science," by Stephen G. Brush; SCIENTIFIC AMERICAN, August 1992].

In 1922 the Russian theorist Alexander A. Friedmann realized that Einstein's universe is unstable; the slightest perturbation would cause it to expand or contract. At that time, Vesto M. Slipher of Lowell Observatory was collecting the first evidence that galaxies are actually moving apart. Then, in 1929, the eminent astronomer Edwin P. Hubble showed that the rate a galaxy is moving away from us is roughly proportional to its distance from us.

The existence of an expanding universe implies that the cosmos has evolved from a dense concentration of matter into the present broadly spread distribution of galaxies. Fred Hoyle, an English cosmologist, was the first to call this process the big bang. Hoyle intended to disparage the theory, but the name was so catchy it gained popularity. It is somewhat misleading, however, to describe the expansion as some type of explosion of matter away from some particular point in space.

That is not the picture at all: in Einstein's universe the concept of space and the distribution of matter are intimately linked; the observed expansion of the system of galaxies reveals the unfolding of space itself. An essential feature of the theory is that the average density in space declines as the universe expands; the distribution of matter forms no observable edge. In an explosion the fastest particles move out into empty space, but in the big bang cosmology, particles uniformly fill all space. The expansion of the universe has had little influence on the size of galaxies or even clusters of galaxies that are bound by gravity; space is simply opening up between them. In this sense, the expansion is similar to a rising loaf of raisin bread. The dough is analogous to space, and the raisins, to clusters of galaxies. As the dough expands, the raisins move apart. Moreover, the speed with which any two raisins move apart is directly and positively related to the amount of dough separating them.

The evidence for the expansion of the universe has been accumulating for some 60 years. The first important clue is the redshift. A galaxy emits or absorbs some wavelengths of light more strongly than others. If the galaxy is moving away from us, these emission and absorption features are shifted to longer wavelengths—that is, they become redder as the recession velocity increases. This phenomenon is known as the redshift.

Hubble's measurements indicated that the redshift of a distant galaxy is greater than that of one closer to the earth. This relation, now known as Hubble's law, is just what one would expect in a uniformly expanding universe. Hubble's law says the recession velocity of a galaxy is equal to its distance multiplied by a quantity called Hubble's constant. The redshift effect in nearby galaxies is relatively subtle, requiring good instrumentation to detect it. In contrast, the redshift of very distant objects—radio galaxies and quasars—is an awesome phenomenon; some appear to be moving away at greater than 90 percent of the speed of light.

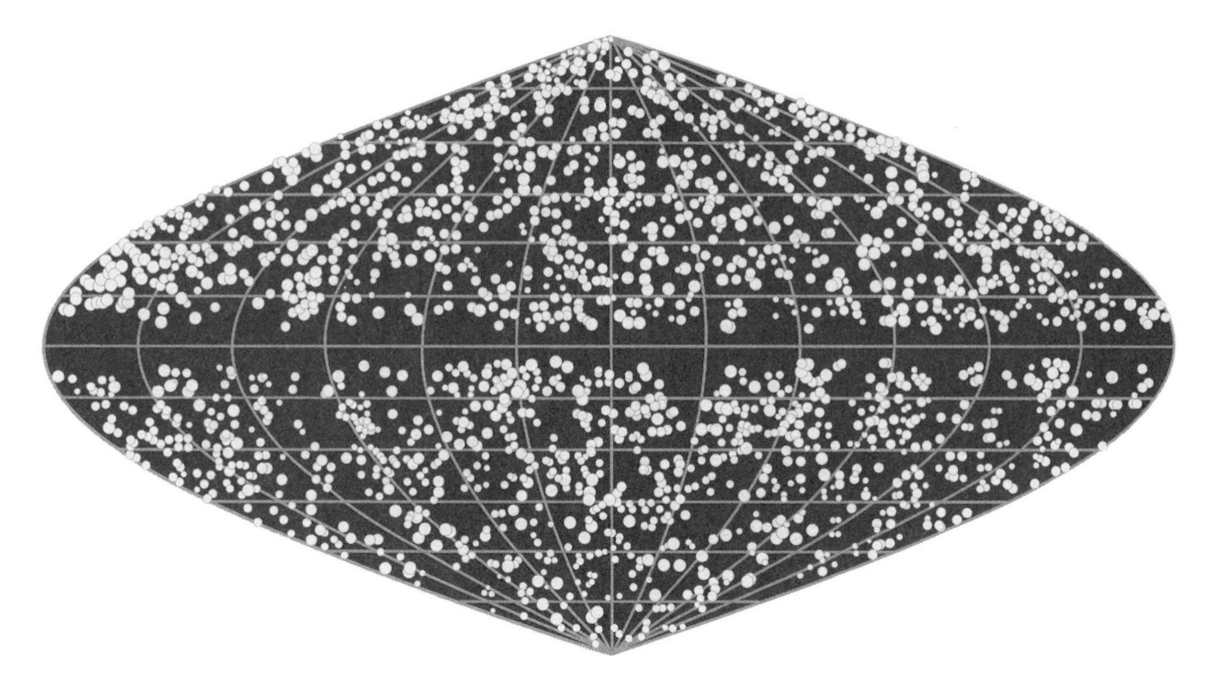

Figure 2.2 HOMOGENEOUS DISTRIBUTION of galaxies is apparent in a map that includes objects from 300 to 1,000 million light-years away. The only inhomogeneity, a gap near the center line, occurs because part of the sky is obscured by the Milky Way. (Michael Strauss of the Institute for Advanced Study in Princeton, N.J., created the map using data from NASA's *Infrared Astronomical Satellite*.)

Hubble contributed to another crucial part of the picture. He counted the number of visible galaxies in different directions in the sky and found that they appear to be rather uniformly distributed. The value of Hubble's constant seemed to be the same in all directions, a necessary consequence of uniform expansion. Modern surveys confirm the fundamental tenet that the universe is homogeneous on large scales. Although maps of the distribution of the nearby galaxies display clumpiness, deeper surveys reveal considerable uniformity (see Figure 2.2).

The Milky Way, for instance, resides in a knot of two dozen galaxies; these in turn are part of a complex of galaxies that protrudes from the so-called local supercluster. The hierarchy of clustering has been traced up to dimensions of about 500 million light-years. The fluctuations in the average density of matter diminish as the scale of the structure being investigated increases. In maps that cover distances that reach close to the observable limit, the average density of matter changes by less than a tenth of a percent.

To test Hubble's law, astronomers need to measure distances to galaxies. One method for gauging distance is to observe the apparent brightness of a galaxy. If one galaxy is four times fainter in the night sky than an otherwise comparable galaxy, then it can be estimated to be twice as far away. This expectation has now been tested over the whole of the visible range of distances.

Some critics of the theory have pointed out that a galaxy that appears to be smaller and fainter might not actually be more distant. Fortunately, there is a direct indication that objects whose redshifts are larger really are more distant. The evidence comes from observations of an effect known as gravitational lensing (see Figure 2.3). An object as massive and compact as a galaxy can act as a crude lens, producing a distorted, magnified image (or even many images) of any background radiation source that lies behind it. Such an object does so by bending the paths of light rays and other electromagnetic radiation. So if a galaxy sits in the line of sight between the earth and some distant object, it will bend the light rays from the object so that they are

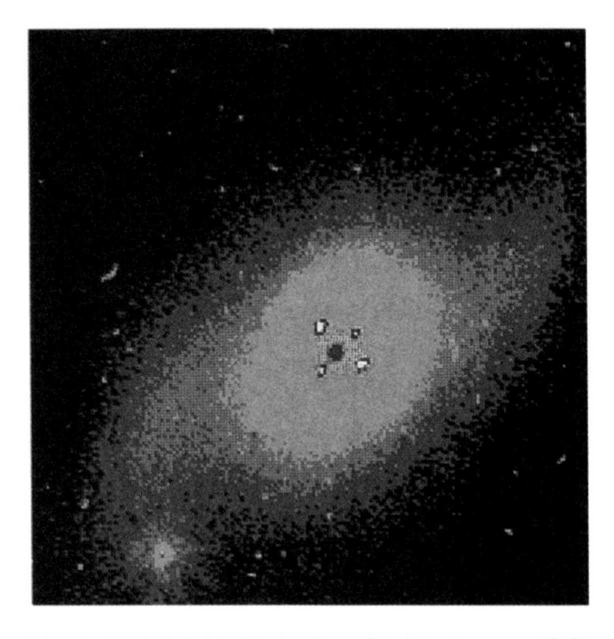

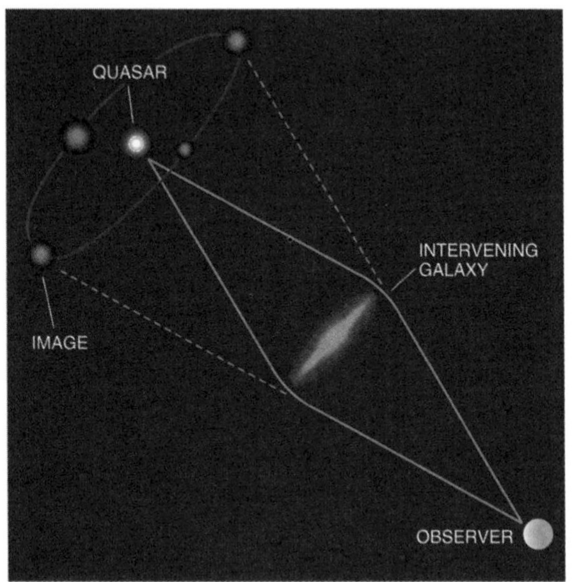

Figure 2.3 MULTIPLE IMAGES of a distant quasar (*left*) are the result of an effect known as gravitational lensing. The effect occurs when light from a distant object is bent by the gravitational field of an intervening galaxy. In this case, the galaxy, which is visible in the center, produces four images of the quasar. (The photograph was produced using the *Hubble* telescope.)

observable [see "Gravitational Lenses," by Edwin L. Turner; Scientific American, July 1988]. During the past decade, astronomers have discovered more than a dozen gravitational lenses. The object behind the lens is always found to have a higher redshift than the lens itself, confirming the qualitative prediction of Hubble's law.

Hubble's law has great significance not only because it describes the expansion of the universe but also because it can be used to calculate the age of the cosmos. To be precise, the time elapsed since the big bang is a function of the present value of Hubble's constant and its rate of change. Astronomers have determined the approximate rate of the expansion, but no one has yet been able to measure the second value precisely.

Still, one can estimate this quantity from knowledge of the universe's average density. One expects that because gravity exerts a force that opposes expansion, galaxies would tend to move apart more slowly now than they did in the past. The rate of change in expansion is therefore related to the gravitational pull of the universe set by its average density. If the density is that of just the visible material in and around galaxies, the age of the universe probably lies between 12 and 20 billion years. (The range allows for the uncertainty in the rate of expansion.)

Yet many researchers believe the density is greater than this minimum value. So-called dark matter would make up the difference. A strongly defended argument holds that the universe is just dense enough that in the remote future the expansion will slow almost to zero. Under this assumption, the age of the universe decreases to the range of seven to 13 billion years.

To improve these estimates, many astronomers are involved in intensive research to measure both the distances to galaxies and the density of the universe. Estimates of the expansion time provide an important test for the big bang model of the universe. If the theory is correct, everything in the visible universe should be younger than the expansion time computed from Hubble's law.

These two timescales do appear to be in at least rough concordance. For example, the oldest stars in the disk of the Milky Way galaxy are about nine billion years old—an estimate derived from the rate of cooling of white dwarf stars. The stars in the halo of the Milky Way are somewhat older, about 15 billion years—a value derived from the rate of nuclear fuel consumption in the cores of these stars.

The ages of the oldest known chemical elements are also approximately 15 billion years—a number that comes from radioactive dating techniques. Workers in laboratories have derived these age estimates from atomic and nuclear physics. It is noteworthy that their results agree, at least approximately, with the age that astronomers have derived by measuring cosmic expansion.

Another theory, the steady state theory, also succeeds in accounting for the expansion and homogeneity of the universe. In 1946 three physicists in England—Hoyle, Hermann Bondi and Thomas Gold—proposed such a cosmology. In their theory the universe is forever expanding, and matter is created spontaneously to fill the voids. As this material accumulates, they suggested, it forms new stars to replace the old. This steady state hypothesis predicts that ensembles of galaxies close to us should look statistically the same as those far away. The big bang cosmology makes a different prediction: if galaxies were all formed long ago, distant galaxies should look younger than those nearby because light from them requires a longer time to reach us. Such galaxies should contain more short-lived stars and more gas out of which future generations of stars will form.

The test is simple conceptually, but it took decades for astronomers to develop detectors sensitive enough to study distant galaxies in detail. When astronomers examine nearby galaxies that are powerful emitters of radio wavelengths, they see, at optical wavelengths, relatively round systems of stars. Distant radio galaxies, on the other hand, appear to have elongated and sometimes irregular structures. Moreover, in most distant radio galaxies, unlike the ones nearby, the distribution of light tends to be aligned with the pattern of the radio emission (see Figure 2.4).

Likewise, when astronomers study the population of massive, dense clusters of galaxies, they find differences between those that are close and those far away. Distant clusters contain bluish galaxies that show evidence of ongoing star formation. Similar clusters that are nearby contain reddish galaxies in which active star formation ceased long ago. Observations made with the *Hubble Space Telescope* confirm that at least some of the enhanced star formation in these younger clusters may be the result of collisions between their member galaxies, a process that is much rarer in the present epoch.

So if galaxies are all moving away from one another and are evolving from earlier forms, it seems logical that they were once crowded together in some dense sea of matter and energy. Indeed, in 1927, before much was known about distant galaxies, a Belgian cosmologist and priest, Georges Lemaître, proposed that the expansion of the universe might be traced to an exceedingly dense state he called the primeval "super-atom." It might even be possible, he thought, to detect remnant radiation from the primeval atom. But what would this radiation signature look like?

When the universe was very young and hot, radiation could not travel very far without being absorbed and emitted by some particle. This continuous exchange of energy maintained a state of thermal equilibrium; any particular region was unlikely to be much hotter or cooler than the average. When matter and energy settle to such a state, the result is a so-called thermal spectrum, where the intensity of radiation at each wavelength is a definite function of the temperature. Hence, radiation originating in the hot big bang is recognizable by its spectrum.

In fact, this thermal cosmic background radiation has been detected. While working on the development of radar in the 1940s, Robert H. Dicke, then at the Massachusetts Institute of Technology, invented the microwave radiometer—a device capable of detecting low levels of radiation. In the 1960s Bell Laboratories used a radiometer in a telescope that would track the early communications satellites *Echo-1* and *Telstar.* The engineer who built this instrument found that it was detecting unexpected radiation. Arno A. Penzias and Robert W. Wilson identified the signal as the cosmic background radiation. It is interesting that Penzias and Wilson were led to this idea by the news that Dicke had suggested that one ought to use a radiometer to search for the cosmic background.

Astronomers have studied this radiation in great detail using the *Cosmic Background Explorer (COBE)* satellite and a number of rocket-launched, balloon-borne and ground-based experiments. The cosmic background radiation has two distinctive properties. First, it is nearly the same in all directions. (As George F. Smoot of Lawrence Berkeley Laboratory and his team discovered in 1992, the variation is just one part per 100,000.) The interpretation is that the radiation uniformly fills space, as predicted in the big bang cosmology. Second, the spectrum is

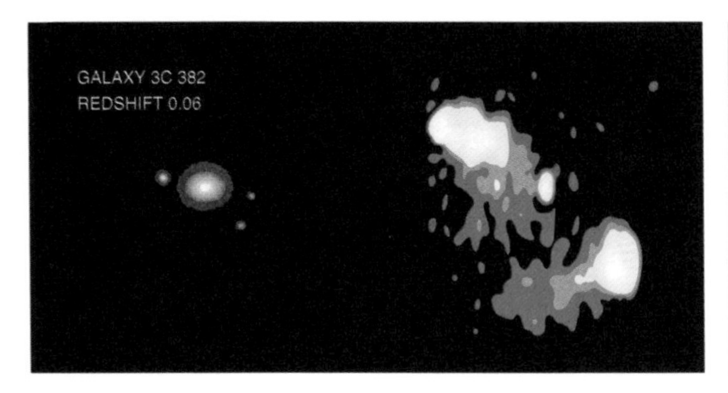

Figure 2.4 DISTANT GALAXIES differ greatly from those nearby—an observation that shows that galaxies evolved from earlier, more irregular forms. Among galaxies that are bright at both optical (*blue*) and radio (*red*) wavelengths, the nearby galaxies tend to have smooth elliptical shapes at optical wavelengths and very elongated radio images. As

very close to that of an object in thermal equilibrium at 2.726 kelvins above absolute zero. To be sure, the cosmic background radiation was produced when the universe was far hotter than 2.726 degrees, yet researchers anticipated correctly that the apparent temperature of the radiation would be low. In the 1930s Richard C. Tolman of the California Institute of Technology showed that the temperature of the cosmic background would diminish because of the universe's expansion.

The cosmic background radiation provides direct evidence that the universe did expand from a dense, hot state, for this is the condition needed to produce the radiation. In the dense, hot early universe thermonuclear reactions produced elements heavier than hydrogen, including deuterium, helium and lithium (see Figure 2.5). It is striking that the computed mix of the light elements agrees with the observed abundances. That is, all evidence indicates that the light elements were produced in the hot, young universe, whereas the heavier elements appeared later, as products of the thermonuclear reactions that power stars.

The theory for the origin of the light elements emerged from the burst of research that followed the end of World War II. George Gamow and graduate student Ralph A. Alpher of George Washington University and Robert Herman of the Johns Hopkins University Applied Physics Laboratory and others used nuclear physics data from the war effort to predict what kind of nuclear processes might have occurred in the early universe and what elements might have been produced. Alpher and Herman also realized that a remnant of the original expansion would still be detectable in the existing universe.

Despite the fact that significant details of this pioneering work were in error, it forged a link between nuclear physics and cosmology. The workers demonstrated that the early universe could be viewed as a type of thermonuclear reactor. As a result, physicists have now precisely calculated the abundances of light elements produced in the big bang and how those quantities have changed because of subsequent events in the interstellar medium and nuclear processes in stars.

Our grasp of the conditions that prevailed in the early universe does not translate into a full understanding of how galaxies formed. Nevertheless, we do have quite a few pieces of the puzzle. Gravity causes the growth of density fluctuations in the distribution of matter, because it more strongly slows the expansion of denser regions, making them grow still denser. This process is observed in the growth of nearby clusters of galaxies, and the galaxies themselves were probably assembled by the same process on a smaller scale.

The growth of structure in the early universe was prevented by radiation pressure, but that changed when the universe had expanded to about 0.1 percent of its present size. At that point, the temperature was about 3,000 kelvins, cool enough to allow the ions and electrons to combine to form neutral hydrogen and helium. The neutral matter was able to slip through the radiation and to form gas clouds that could collapse to star clusters. Observations show that by the time the universe

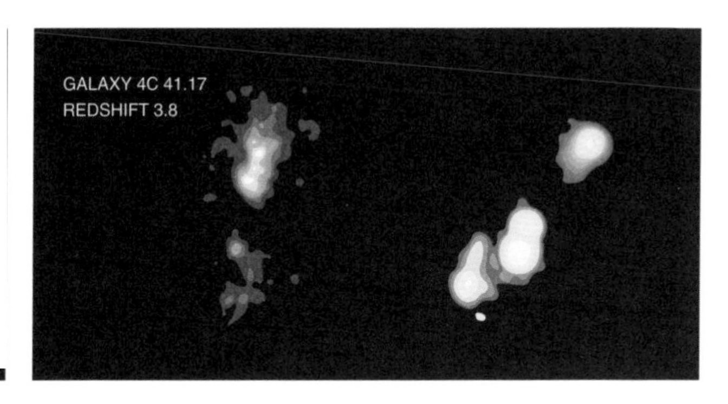

redshift, and therefore distance, increases, galaxies have more irregular elongated forms that appear aligned at optical and radio wavelengths. The galaxy at the far right is seen as it was at 10 percent of the present age of the universe. (The images were assembled by Pat McCarthy of the Carnegie Institute.)

was one fifth its present size, matter had gathered into gas clouds large enough to be called young galaxies.

A pressing challenge now is to reconcile the apparent uniformity of the early universe with the lumpy distribution of galaxies in the present universe. Astronomers know that the density of the early universe did not vary by much, because they observe only slight irregularities in the cosmic background radiation. So far it has been easy to develop theories that are consistent with the available measurements, but more critical tests are in progress. In particular, different theories for galaxy formation predict quite different fluctuations in the cosmic background radiation on angular scales less than about one degree. Measurements of such tiny fluctuations have not yet been done, but they might be accomplished in the generation of experiments now under way. It will be exciting to learn whether any of the theories of galaxy formation now under consideration survive these tests.

The present-day universe has provided ample opportunity for the development of life as we know it—there are some 100 billion billion stars similar to the sun in the part of the universe we can observe. The big bang cosmology implies, however, that life is possible only for a bounded span of time: the universe was too hot in the distant past, and it has limited resources for the future. Most galaxies are still producing new stars, but many others have already exhausted their supply of gas. Thirty billion years from now, galaxies will be much darker and filled with dead or dying stars, so there will be far fewer planets capable of supporting life as it now exists.

The universe may expand forever, in which case all the galaxies and stars will eventually grow dark

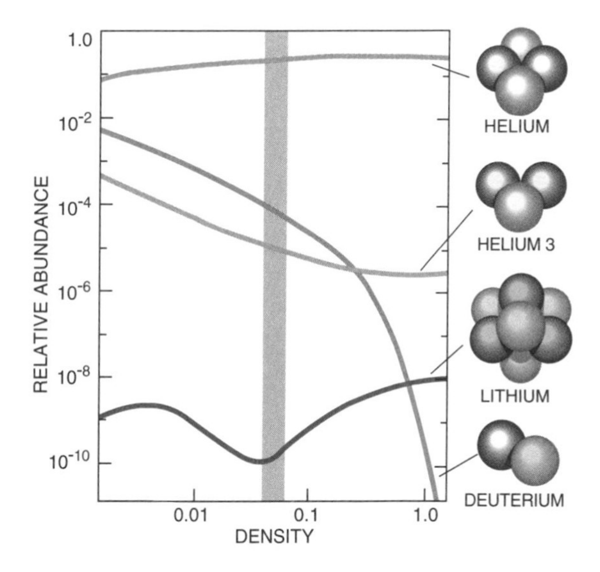

Figure 2.5 DENSITY of neutrons and protons in the universe determined the abundances of certain elements. For a higher density universe, the computed helium abundance is little different, and the computed abundance of deuterium is considerably lower. The shaded region is consistent with the observations, ranging from an abundance of 24 percent for helium to one part in 10^{10} for the lithium isotope. This quantitative agreement is a prime success of the big bang cosmology.

and cold. The alternative to this big chill is a big crunch. If the mass of the universe is large enough, gravity will eventually reverse the expansion, and all matter and energy will be reunited. During the next decade, as researchers improve techniques for measuring the mass of the universe, we may learn whether the present expansion is headed toward a big chill or a big crunch.

In the near future, we expect new experiments to provide a better understanding of the big bang. As we improve measurements of the expansion rate and the ages of stars, we may be able to confirm that the stars are indeed younger than the expanding universe. The larger telescopes recently completed or under construction may allow us to see how the mass of the universe affects the curvature of space-time, which in turn influences our observations of distant galaxies.

We will also continue to study issues that the big bang cosmology does not address. We do not know why there was a big bang or what may have existed before. We do not know whether our universe has siblings—other expanding regions well removed from what we can observe. We do not understand why the fundamental constants of nature have the values they do. Advances in particle physics suggest some interesting ways these questions might be answered; the challenge is to find experimental tests of the ideas.

In following the debate on such matters of cosmology, one should bear in mind that all physical theories are approximations of reality that can fail if pushed too far. Physical science advances by incorporating earlier theories that are experimentally supported into larger, more encompassing frameworks. The big bang theory is supported by a wealth of evidence: it explains the cosmic background radiation, the abundances of light elements and the Hubble expansion. Thus, any new cosmology surely will include the big bang picture. Whatever developments the coming decades may bring, cosmology has moved from a branch of philosophy to a physical science where hypotheses meet the test of observation and experiment.

The Earth's Elements

The elements that make up
the earth and its inhabitants were created
by earlier generations of stars.

• • •

Robert P. Kirshner

Matter in the universe was born in violence. Hydrogen and helium emerged from the intense heat of the big bang some 15 billion years ago. More elaborate atoms of carbon, oxygen, calcium and iron, out of which we are made, had their origins in the burning depths of stars. Heavy elements such as uranium were synthesized in the shock waves of supernova explosions. The nuclear processes that created these ingredients of life took place in the most inhospitable of environments.

Once formed, violent explosions returned the elements to the space between the stars. There gravitation molded them into new stars and planets, and electromagnetism cast them into the chemicals of life. The ink on this page, the air you breathe while reading it—to say nothing of your bones and blood—are all an inheritance from earlier generations of stars. Walking down the corridors of an observatory, you see collections of carbon atoms hunched over silicon boxes, controlling distant telescopes of iron and aluminum in an attempt to trace the origin of the very substances of which they are made.

Matter was created in a violent explosion, known as the big bang, some 15 billion years ago. Within a minute fraction of a second, newborn quarks coalesced into protons. These fused further into the nuclei of helium atoms. Gravitational forces amplified ripples in this primordial soup, pulling the densest regions together into a giant cosmic tapestry of galaxies and voids. Inside galaxies, thick clouds of gas spawned stars. Traces of those early ripples can be seen in the cosmic microwave radiation, which still bears traces of the structure in the infant universe.

The large-scale unfolding of the universe was accompanied by a parallel change in the microscopic structure of matter. Carbon and nitrogen and other elements essential to life on the earth were synthesized in the interiors of stars now long deceased (see Figure 3.1). Within the Milky Way galaxy, in the familiar stars of the night sky, astronomers can study these processes of microscopic change. In the early 1900s, such studies led to the first of several paradoxes regarding the ages of planets and stars.

The study of natural radioactivity on the earth provided clues about the ages of the elements. Geophysicists looking at the slow decay of uranium into lead computed an age for the earth of a few billion years. But astrophysicists of the early 20th century, not knowing about nuclear processes, computed that a sun powered by chemical burning

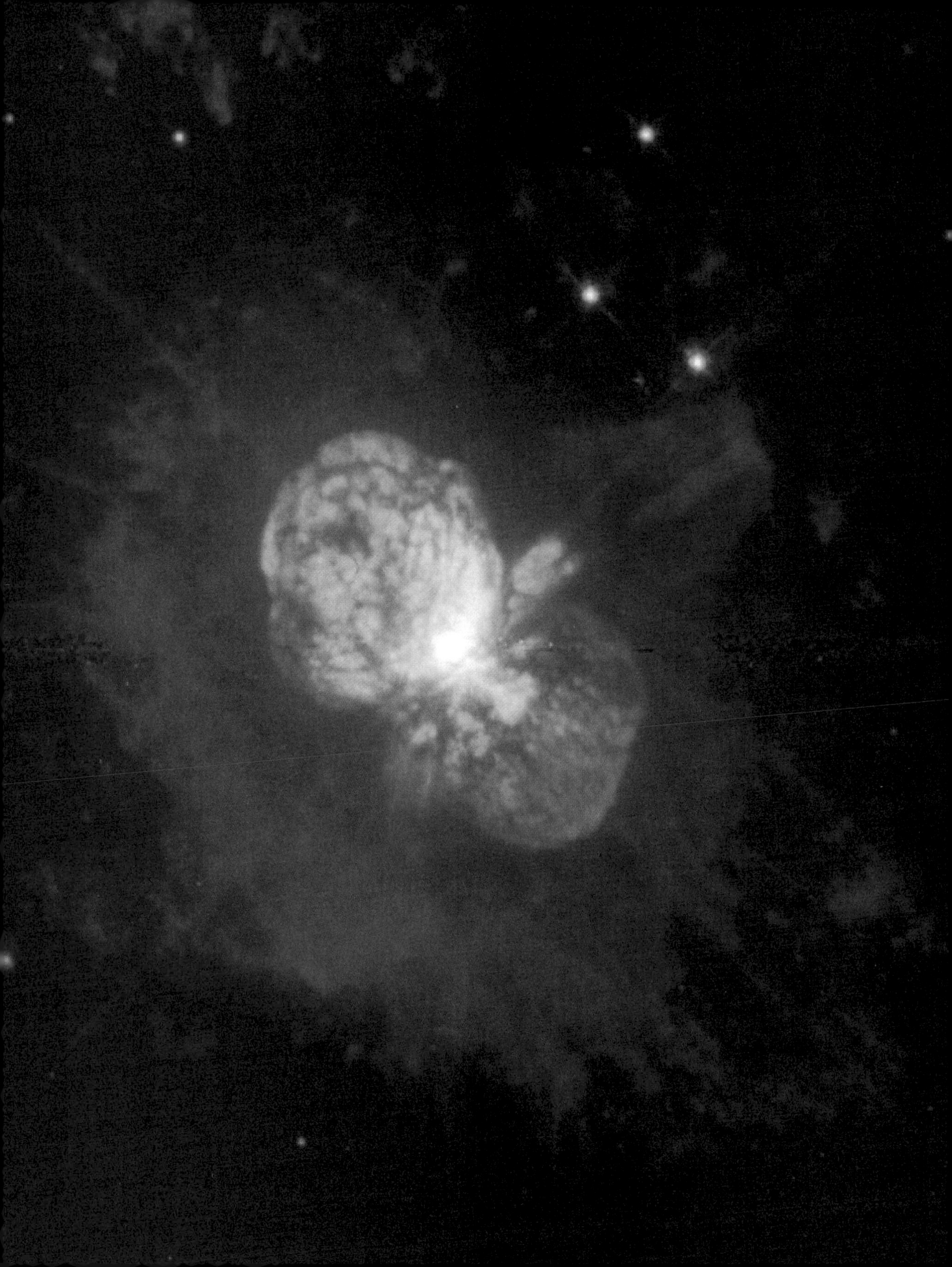

or gravitational shrinking could shine only for a few million years.

The discrepancy mattered. An age of billions of years for the earth provides a much more plausible calendar for biological and geologic evolution, where humans often find that change is imperceptibly slow. Even though the rug in most astronomy departments is lumpy from all the discrepancies that have been swept under it, a factor of 1,000 demands attention.

Curiously, the key to the problem was found in the processes of nuclear physics that, in the form of radioactivity, had first posed it. If stars live for billions of years instead of millions, they must have a continuing source of energy 1,000 times larger than chemical energy. Ordinary chemical changes involve the electrical force rearranging electrons in the outer regions of atoms. Nuclear changes involve the strong force rearranging neutrons and protons within the nucleus of an atom. The products of the reaction sometimes have less mass than the ingredients; the excess mass is converted to energy according to the well-known formula $E = mc^2$.

In nuclear reactions the energy yield is extremely large, typically a million times the energy produced by chemical reactions. Even the terminology for nuclear weapons reflects this factor. The unit of nuclear energy is a megaton—the energy of a million tons of chemical explosive.

A star that burns hydrogen, such as the sun, has an ample supply of energy for a lifetime of 10 billion years. Estimates for the current age of the sun are in the vicinity of five billion years (so we can safely contract for long-term mortgages).

The nuclear reactions within stars provide more than the energy that allows life to flourish. The ashes of nuclear burning—the elements of the periodic table—are the materials out of which living things are made. Perhaps most important, nuclear fusion, occurring steadily over the lifetime of a star, ensures a continuous supply of energy for billions of years and allows time for life and intelligence to develop (see Figure 3.2).

Figure 3.1 ETA CARINAE, a star thought to be of 150 solar masses more than 10,000 light-years away, had a violent outburst in 1841. The *Hubble Space Telescope* image reveals two plumes, made of nitrogen and other elements synthesized in the interior of the star, moving out into the interstellar void at more than two million miles per hour. Some elements making up the earth came from similar discharges from ancestral stars.

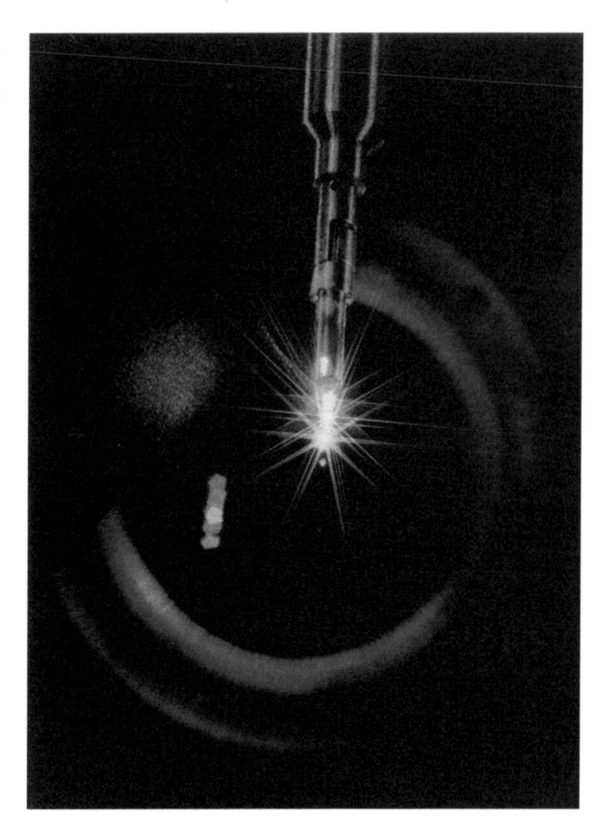

Figure 3.2 CAPTIVE STAR is created when Lawrence Livermore National Laboratory's Nova laser beams implode a capsule containing deuterium and tritium. Ten symmetrically arranged laser tubes, one of which is seen head-on (*red circle*), shine more than 100 trillion watts of power onto the capsule mounted at the tip of the vertical assembly. The capsule collapses, compressing the atoms inside to sufficiently high temperature and density that fusion takes place. Such artificial suns, it is hoped, will one day meet the energy needs of humankind.

Stars, after all, are not such ordinary places in the universe. A star is a ball of gas neatly balanced between the inward pull of its own gravitation and the outward pressure of the hot gas within. The compressed hydrogen gas usually has the density of the water in Boston Harbor, some 10^{30} times higher than the norm in the universe. And in a universe with a typical temperature of three kelvins (–270 degrees Celsius), the center of a star is at 15 million kelvins.

At such extreme temperatures the hydrogen atoms are stripped of their electrons. The naked protons undergo frequent, jarring collisions as they buzz furiously in the star's dense interior. Near the center the temperature and density are highest. There the protons, despite the electrical repulsion between them, are pushed so close together that the strong and the weak nuclear forces can come into play.

In a series of nuclear reactions, hydrogen nuclei (protons) fuse into helium nuclei (two protons and two neutrons), emitting two positrons, two neutrinos and energy. If the elements synthesized were limited to helium (which is also made in the big bang) and if it stayed locked up in the cores of stars, this would not be quite such an interesting story—and we would not be here to discuss it. After a long and steady phase of hydrogen fusion, which leads to helium accumulating in the core, the star changes dramatically.

The core shrinks and heats as four nucleons are locked up in each helium nucleus. The temperature and density of the core increase to maintain the pressure balance. The star as a whole becomes less homogeneous. While the core becomes smaller, the outer layers swell up to 50 times their previous radius. A star the size of the sun will swiftly transform into a cool, but luminous, red giant. From the parochial viewpoint of earth dwellers, this will be the end of history and of human creations. Commodity future options, the designated-hitter rule and call waiting will all be vaporized with the earth.

But interesting events take place inside red giants. As the core contracts, the central furnace grows denser and hotter. Then nuclear reactions that were previously impossible become the principal source of energy. For example, the helium that accumulates during hydrogen burning can now become a fuel. As the star ages and the core temperature rises, brief encounters between helium nuclei produce fusion events.

The collision of two helium nuclei leads initially to an evanescent form of beryllium having four neutrons and four protons. Amazingly enough, another helium nucleus collides with this short-lived target, leading to the formation of carbon. The process would seem about as likely as crossing a stream by stepping fleetingly on a log. A delicate match between the energies of helium, the unstable beryllium and the resulting carbon allows the last to be created. Without this process, we would not be here.

Carbon and oxygen, formed by fusing one more helium with carbon, are the most abundant elements formed in stars. The many collisions of protons with helium atoms do not give rise to significant fusion products. Lithium, beryllium and boron—the nuclei of which are smaller than those of carbon—are a million times less abundant than carbon. Thus, abundances of elements are determined by often obscure details of nuclear physics. A star of the sun's mass endures as a red giant for only a few hundred million years. The last stages of burning are unstable: the star pushes off its outer layers to form a shell of gas called a planetary nebula. In some stars, carbon-rich matter from the core is dredged up by convection. The freshly synthesized matter then escapes, forming a sooty cocoon of graphite. Eventually fuel runs out, and the inner core of the red giant congeals into a white dwarf.

A white dwarf is protected from total gravitational collapse not by the kinetic pressure of gases; the carbon and oxygen in its interior are in an almost crystalline state. The star is held up by the quantum repulsion of its free electrons. Quantum mechanics forbids electrons from sharing the lowest energy state. This restriction forces most electrons to occupy higher energy states even though the gas is relatively cold. These electrons provide the pressure to support a white dwarf. There is no more generation of nuclear energy, and no new elements are synthesized.

Many white dwarfs in our galaxy come to this dull end, slowly cooling, dimming and slipping below the edge of detection. Sometimes a too generous neighboring star may supply gas that streams onto a white dwarf, provoking it into a type I supernova and a sudden synthesis of new elements.

The most significant locations for the natural alchemy of fusion are, however, stars more massive than the sun. Although rarer, a heavy star follows a shorter and more intense path to destruction. To support the weight of the star's massive outer layers, the temperature and pressure in its core have to be high. A star of 20 solar masses is more than 20,000 times as luminous as the sun. Rushing through its hydrogen-fusion phase 1,000 times faster, it swells up to become a red giant in just 10 million years instead of the sun's 10 billion.

The high central temperature leads as well to a more diverse set of nuclear reactions. A sunlike star builds up carbon and oxygen that stays locked in the cooling ember of a white dwarf. Inside a massive star, carbon nuclei fuse further to make neon and magnesium. Fusion of oxygen yields silicon as well, along with sulfur. Silicon burns to make iron. Intermediate stages of fusion and decay make many different elements, all the way up to iron.

The iron nucleus occupies a special place in nuclear physics and, by extension, in the composition of the universe. Iron is the most tightly bound nucleus. Lighter nuclei, when fusing together, release energy. To make a nucleus heavier than iron, however, requires an expenditure of energy. This fact, established in terrestrial laboratories, is instrumental in the violent death of stars. Once a star has built an iron core, there is no way it can generate energy by fusion. The star, radiating energy at a prodigious rate, becomes like a teenager with a credit card. Using resources much faster than can be replenished, it is perched on the edge of disaster.

So what happens? For the star, at least, the disaster takes the form of a supernova explosion. The core collapses inward in just one second to become a neutron star or black hole. The material in the core is as dense as that within a nucleus. The core can be compressed no further. When even more material falls into this hard core, it rebounds like a train hitting a wall. A wave of intense pressure traveling faster than sound—a sonic boom—thunders across the extent of the star. When the shock wave reaches the surface, the star suddenly brightens and explodes. For a few weeks, the surface shines as brightly as a billion suns while the emitting surface expands at several thousand kilometers per second. The abrupt energy release is comparable to the total energy output of the sun in its entire lifetime.

Such type II supernova explosions play a special role in the chemical enrichment of the universe (see Figure 3.3). First, unlike stars of low mass that lock up their products in white dwarfs, exploding stars

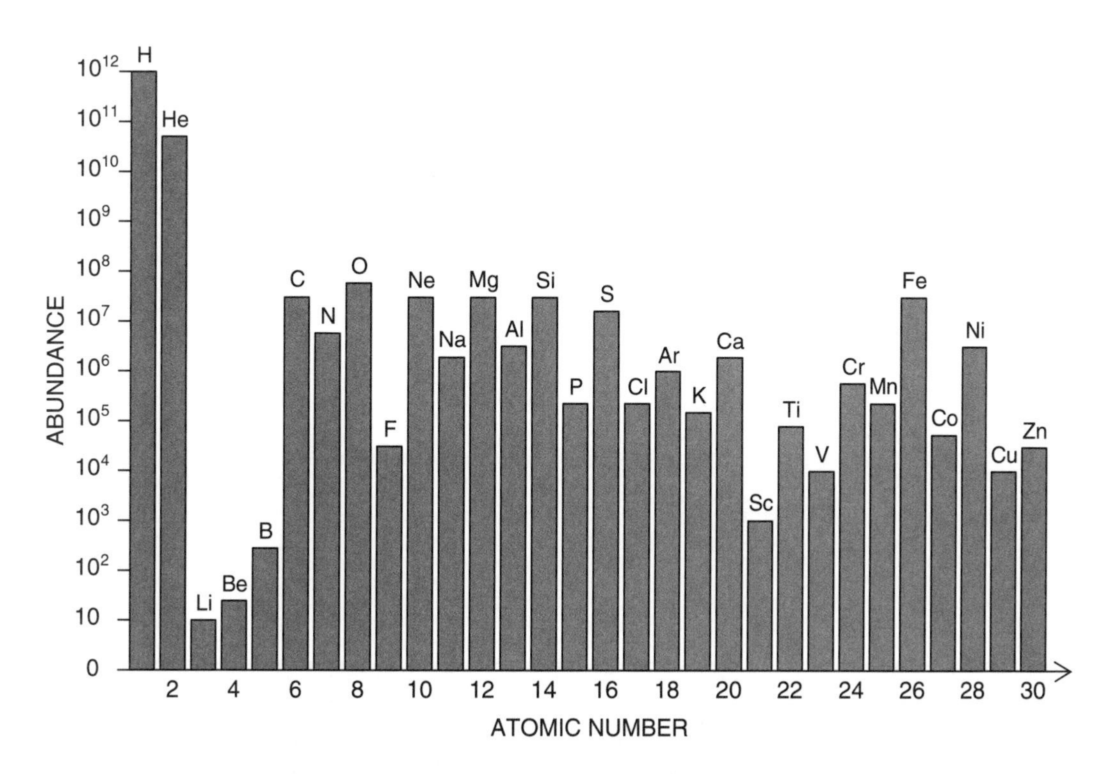

Figure 3.3 RELATIVE ABUNDANCES OF ELEMENTS in the universe reveal the processes that synthesized heavier elements out of the hydrogen (H) and helium (He) of the big bang. Fusion in stars created more helium, skipped over lithium (Li), beryllium (Be) and boron (B) to carbon (C) and generated all the elements up to iron (Fe). Massive stars can synthesize elements heavier than oxygen (O); these stars eventually explode as supernovae. Elements heavier than iron are made in such explosions. The chart has a logarithmic scale, in which abundance increases by a factor of 10 for each unit of height.

eject their outer layers, which are unburned. They belch out the helium that was formed from hydrogen burning and launch the carbon, oxygen, sulfur and silicon that have accumulated from further burning into the gas in their neighborhood.

New elements are synthesized behind the outgoing shock wave. The intense heat enables nuclear reactions that cannot occur in steadily burning stars. Some of the nuclear products are radioactive, but stable elements heavier than iron can also be synthesized. Neutrons bombard iron nuclei, forging them into gold. Gold is transformed into lead (an alchemist's nightmare!), and lead is bombarded to make elements all the way up to uranium. Elements beyond iron in the periodic table are rare in the cosmos. For every 100 billion hydrogen atoms, there is one uranium atom—each made at special expense in an uncommon setting.

This theoretical picture of the creation of heavy elements in supernova explosions was thoroughly tested in February 1987. A supernova, SN 1987A, exploded in the nearby Large Magellanic Cloud (see boxed figure "Supernova 1987A and the Age of the Universe"). Sanduleak –69° 202, which in 1986 was noted as a star of 20 solar masses, is no longer there. Together the star and the supernova give dramatic evidence that at least one massive star ended its life in a violent way.

Neutrinos emitted from the innermost shock wave of the explosion were detected in Ohio and in Japan, hours before the star began to brighten. Freshly synthesized elements radiated energy, making the supernova debris bright enough to see with the naked eye for months after the explosion. In addition, satellites and balloons detected the specific high-energy gamma rays that newborn radioactive nuclei emit.

Observations made in 1987 with the *International Ultraviolet Explorer* and subsequently with the *Hubble Space Telescope* supply strong evidence that Sanduleak –69° 202 was once a red giant star that shed some of its outer layers. Images taken in 1994 with the newly acute *Hubble* revealed astonishing rings around the supernova.

The inner ring is material that the star lost when it was a red giant, excited by the flash of ultraviolet light from the supernova. The outer rings are more mysterious but are presumably related to mass lost from the pre-supernova system. The products of stellar burning are concentrated in a central dot, barely resolved with the *Hubble* telescope, which is expanding outward at 3,000 kilometers per second. No neutron star has yet been observed in SN 1987A.

The supernova has provided dramatic confirmation of elaborate theoretical models of the origin of elements. Successive cycles of star formation and destruction enrich the interstellar medium with heavy elements. We can identify the substances in interstellar gas: they absorb particular wavelengths of light from more distant sources, leaving a characteristic imprint (see Figure 3.4). The absorption lines tell us as well the abundance of the element—its amount compared with that of hydrogen.

In a spiral galaxy like the Milky Way, interstellar gas is associated with the spiral arms. Optical studies of the galaxy are hampered by the accompanying dust, which absorbs much of the light passing through. But the dust also shields the hydrogen atoms from ultraviolet light, allowing them to combine chemically and form molecules (H_2). In these hidden backwaters of the galaxy, other molecules such as water (H_2O), carbon monoxide (CO) and ammonia (NH_3) all assemble. The chemical variety is quite surprising: more than 100 molecules have been found in interstellar clouds.

In May 1994 Yanti Miao and Yi-Jehng Khan of the University of Illinois reported finding the smallest amino acid, glycine, in the star-forming

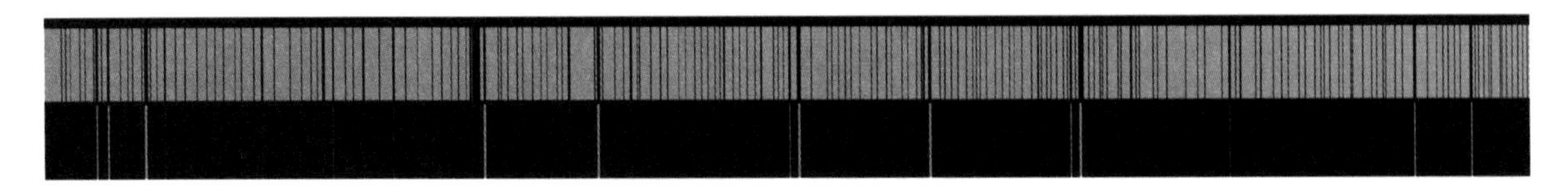

Figure 3.4 SPECTRUM OF THE SUN shows dark absorption lines that coincide with the bright lines in the spectrum of iron (*bottom*). Cool iron atoms absorb the same wavelengths of light that iron atoms emit when hot. The matching lines prove that the sun's relatively cool surface, or photosphere, contains iron, which could have come only from an ancestral star.

Supernova 1987A and the Age of the Universe

Supernova 1987A led to an unexpected, and stringent, test of our ability to measure cosmic distances. Remote stars and galaxies appear to be moving away from the earth, sharing the cosmic expansion that began with the big bang. If we can measure the distance to a receding galaxy, then by combining this information with how fast the galaxy is moving, we can determine for how long it has been receding. Thus, we gain a measure of the age of the universe.

Based on observations we had carried out in 1987 and 1988, my colleagues and I could time how long light took to reach the supernova's bright inner ring. Because we know the speed of light, that time allowed us to calculate the ring's physical size. Observations made with the imperfect *Hubble Space Telescope* in 1990 gave a measure of the ring's apparent angular size, viewed from the solar system. Combining these two pieces of information yields a distance to the Large Magellanic Cloud (in which SN 1987A occurred) of about 169,000 light-years, in good agreement with classical methods.

A separate method we developed to measure the distance to SN 1987A analyzes the light emitted from the supernova shortly after the explosion. When the shock wave reached the surface, it heated the gas and blasted it outward. The velocity with which this debris is flying out is coded in the amount by which the absorption lines of known elements is shifted. Knowing this velocity and the time when the supernova exploded, we can compute how far the debris must have traveled—and therefore the current radius of the supernova. Given the radius, we know its surface area.

A key piece of information now comes into play. From the overall color of the gas we can estimate the supernova's temperature. The latter yields the amount of light the supernova is emitting per unit area of its surface. Because we know the surface area, we can find the total amount of energy being radiated. Measuring the amount of energy received at the earth, we acquire another estimate of how far away SN 1987A is. In repeated calculations of this kind, we get a distance of about 160,000 light-years—an excellent match with the previous estimate by astronomical standards.

With the confidence that this second method gives the "right" answer when used nearby, we have applied it to more distant supernova explosions. My students Ronald Eastman and Brian Schmidt and I have now measured a dozen supernova distances. When combined with the redshifts of the galaxies in which they erupted, the distances yield an age for the universe of between 12 and 16 billion years.

The estimate assumes that gravity has not slowed down the expansion significantly. Many cosmologists suspect that the universe has just enough mass to balance the energy of expansion, slowing it down until it almost stops. If this is so, the age of the universe would be only two thirds the original estimate, which assumed constant expansion. Then the age of the universe should be scaled back to between eight and 11 billion years.

Globular clusters, on the other hand, are between 12 and 18 billion years old. When future measurements determine the deceleration of the universe, I expect they will do so in the direction of avoiding a paradox. It would be embarrassing to find 14-billion-year-old globular clusters in a universe that is aged only seven billion years.

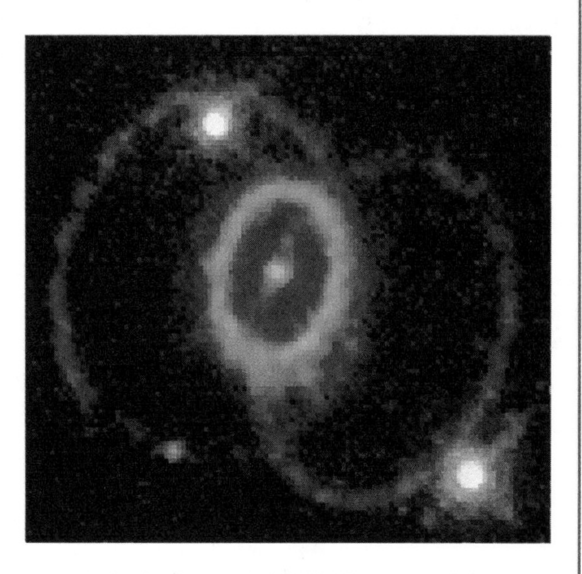

BRIGHT RINGS around SN 1987A are material emitted early in the star's life, heated by light from the explosion.

cloud near the center of our galaxy, Sagittarius B2. It is amusing to speculate that amino acids and other biologically interesting chemicals could be present in the protoplanetary disk that accumulates near a forming star. Such chemicals, if on a young planet, would almost certainly be destroyed by heat. But after the planet had cooled, they could reach its surface by way of meteorites. Indeed, complex hydrocarbons were found last year on microscopic dust particles that originated in interplanetary space.

We can learn much about the materials from which the earth was formed by the simple act of picking up a pen. Made of carbon compounds and metals, the pen—and indeed the earth itself—is typical of the cosmic pattern of abundances. Except for hydrogen and helium, which easily slip the gravitational grip of a small planet, the elements of the earth are the elements of the universe: formed by stars and dispersed throughout the galaxy. (The jury is still out on the question of whether ordinary matter, composed of known subatomic particles, is a small fraction of the total mass in the universe. If so, then we are truly made of uncommon stuff.)

Whereas the sun is 99 percent hydrogen and helium, the 1 percent of more complex nuclei includes traces of iron and other heavy elements. Thus, the solar system must have formed from elements synthesized by previous generations of stars. Like silver candlesticks from your grandmother (but much more valuable), we have inherited the carbon and oxygen produced by ancestral stars.

Astronomers can begin to trace a family tree for the solar system by examining massive stars within the Milky Way. If the massive stars in a star cluster are just now becoming red giants, the cluster must be young. If the stars currently headed toward the red giant phase have the mass of the sun, the cluster must be old enough for its sunlike stars to begin that change: about 10 billion years. The oldest clusters in our galaxy are the globular clusters, which appear to have an age of 12 to 18 billion years when measured in this way.

We recognize the globular clusters as an early generation of stars. The oldest of these are significantly different from the sun; the abundances of elements such as iron are often 100 or even 1,000 times lower. Yet even these ancient stars contain a pinch of heavy elements. Thus, they evince the presence of a completely unseen generation of stars, which has no members left.

Given that the universe itself is only about 15 billion years old, the initial chemical enrichment of the Milky Way must have been very rapid. (Even quasars, extragalactic beacons from a time when the universe was only a fifth of its current age, contain carbon and nitrogen.) There has been much less change in recent times. The present-day chemical abundances in interstellar gas are about the same as in the sun, locked in five billion years ago. This is the raw material for future stars and planets.

In neighboring gas clouds such as the Orion nebula, astronomers can study intimate scenes of stellar birth. New infrared detectors are lifting the shroud from these cradles. (Although it blocks visible light, interstellar dust is transparent to infrared or radio waves.) We can see infant stars as they condense, even before they ignite hydrogen fuel in their cores (see Figure 3.5). In addition, large telescopes such as the eight-meter Gemini telescopes in Hawaii and Chile promise much more detail about the process by which stars condense.

As gas coalesces into a star, it first forms a rotating disk of gas and dust. While the star condenses, the dust aggregates into rocky planets, such as the earth. Residual gas accumulates to make large gas planets, such as Jupiter. Disks, observed with infrared and radio techniques and, occasionally, glimpsed with optical methods, are common. Are planets?

The evidence is much weaker than the conviction. As in cosmology, where there is one example of a universe (we are in it), there is one well-known planetary system (we are on it). A planet is difficult to sight directly. An observer would have to see a small object, shining only by reflected light, next to one about a billion times brighter.

Detecting planets by their gravitational effects is more promising. The idea is to observe the velocity changes of a visible star produced by an unseen object as the two execute a stellar do-si-do. The object, having less than a tenth of the mass of the star, would affect the motion of the star only minutely. Although there are tantalizing hints, no planet has yet been discovered by looking for the motion it produces in the luminous star it orbits. Present techniques are not quite up to the task of detecting a planet smaller than Jupiter in orbit around a star like the sun.

Yet a spinning neutron star, PSR B1257+12, was recently shown to have objects that are producing periodic shifts in its emission (see Figure 3.6).

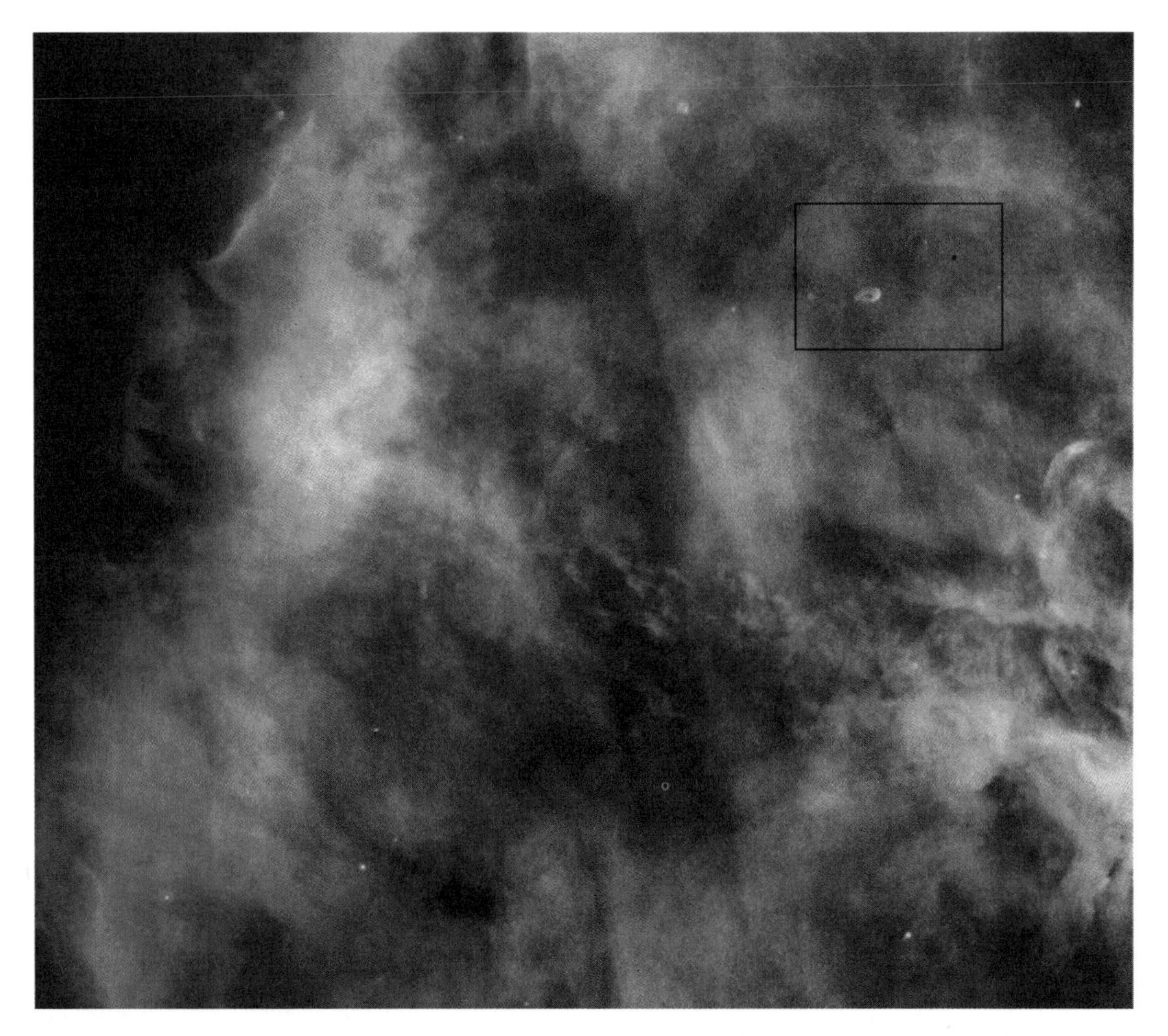

Figure 3.5 STAR CRADLE is found in the Great Nebula in Orion, 1,500 light-years away (*above*). This picture from the *Hubble Space Telescope* codes the presence of nitrogen (*red*) and oxygen (*blue*). At least half the young stars are surrounded by disks of gas and dust from which young planets are believed to form. The magnified image of the outlined part above shows four young stars (*left*). Protoplanetary disks that are lit by hot stars are bright. The cool star, shown magnified (*right*), has one fifth the mass of the sun; its disk contains seven times the material of the earth.

When a neutron star forms in a supernova explosion, the core of the star contracts to a dense sphere just a few miles across. As it shrinks, any rotation of the original star ends up in the rotation of the neutron star. So neutron stars are born spinning. If the neutron star has a magnetic field, it may be a powerful source of radio waves, emitted in a sharply specific direction.

These objects actually exist: they are called pulsars. Every time the fan of radio emission sweeps by the earth, astronomers observe a pulse of radio noise. Because the emission mechanism is anchored to a dense flywheel, the pulse period is very precise. Extremely subtle variations can be measured by diligently observing the arrival times of the pulse. If the pulsar has an unseen companion, an observer will see the pulses arrive a little early, and then late, as the source approaches and recedes.

In 1992 Alexander Wolszczan, now at Pennsylvania State University, and Dale A. Frail of the

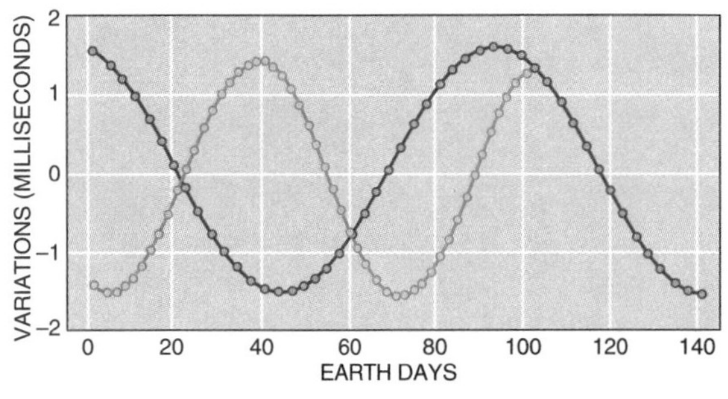

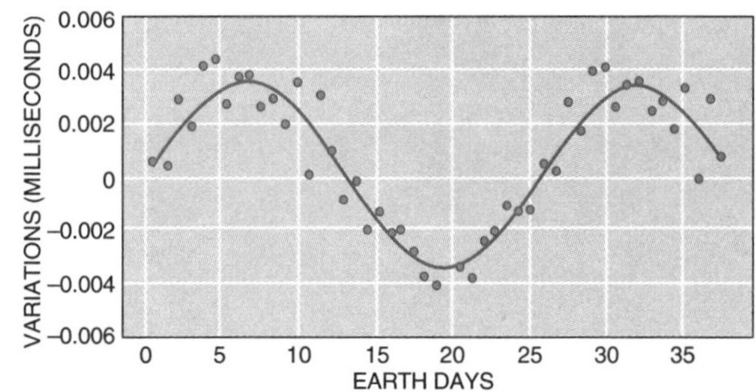

Figure 3.6 PULSAR PSR B1257+12 has at least three planets in orbit around it, the only planets known outside the solar system. The pulsar moves to and fro as the planets orbit it; its pulses reach the earth sometimes sooner, sometimes later, thus revealing the presence of the planets. The graphs show three variations in the times at which the radiation from the pulsar arrived at the earth. The first two variations (*left*) are large, attesting to planets about three times as massive as the earth, with orbital periods of 66.6 (*green*) and 98.2 earth days (*purple*), respectively. The third planet is very close to the pulsar but produces a small variation (*orange*). It has a hundredth the mass of the earth, and its year is just 25.3 days.

National Radio Astronomy Observatory in Socorro, N.M., reported that their observations of the pulsar PSR B1257+12 had periodic changes in the pulse arrival times. The variation was only 1.5 milliseconds, stretched over months. It could be explained if the neutron star was being orbited by a pair of objects. These would have masses of 3.4 and 2.8 times the mass of the earth. In April 1994 these workers found signs of the gravitational forces between the planets and evidence for yet a third object, having about the mass of the moon.

A spinning remnant of a supernova explosion, beaming out powerful radio blasts, is nobody's vision of another solar system. Yet only a curmudgeon could fail to call its orbiting objects planets. It seems quite unlikely that these planets survived the supernova explosion that created the neutron star. The original star probably had a close binary companion, which is no longer present. The planets are perhaps formed from shreds of the companion. This is not your ordinary family history. Nevertheless, the study of pulsars may well shed light on the formation of more familiar planets such as the earth.

The composition of the earth is the natural by-product of energy generation in stars and successive waves of stellar birth and death in our galaxy. We do not know if other stars have earthlike planets where complex atoms, formed in stellar cauldrons, have organized themselves into intelligent systems. But understanding the history of matter and searching for its most interesting forms, such as galaxies, stars, planets and life, seem a suitable use for our intelligence.

The Evolution of the Earth

The formation of this planet and its atmosphere gave rise to life, which shaped the earth's subsequent development. Our future lies in interpreting this geologic past.

• • •

Claude J. Allègre and Stephen H. Schneider

Like the lapis lazuli gem it resembles, the blue, cloud-enveloped planet that we recognize immediately from satellite pictures seems remarkably stable. Continents and oceans, encircled by an oxygen-rich atmosphere, support familiar life-forms. Yet this constancy is an illusion produced by the human experience of time. The earth and its atmosphere are continuously altered. Plate tectonics shift the continents, raise mountains and move the ocean floor while processes that no one fully comprehends alter the climate.

Such constant change has characterized the earth since its beginning some 4.5 billion years ago (see Figure 4.1). From the outset, heat and gravity shaped the evolution of the planet. These forces were gradually joined by the global effects of the emergence of life. Exploring this past offers us the only possibility of understanding the origin of life and, perhaps, its future.

Scientists used to believe the rocky planets, including the earth, Mercury, Venus and Mars, were created by the rapid gravitational collapse of a dust cloud, a deflation giving rise to a dense orb. In the 1960s the Apollo space program changed this view. Studies of moon craters revealed that these gouges were caused by the impact of objects that were in great abundance about 4.5 billion years ago. Thereafter, the number of impacts appeared to have quickly decreased. This observation rejuvenated the theory of accretion postulated by Otto Schmidt. The Russian geophysicist had suggested in 1944 that planets grew in size gradually, step by step.

According to Schmidt, cosmic dust lumped together to form particulates, particulates became gravel, gravel became small balls, then big balls, then tiny planets, or planetesimals, and, finally, dust became the size of the moon. As the planetesimals became larger, their numbers decreased. Consequently, the number of collisions between planetesimals, or meteorites, decreased. Fewer items available for accretion meant that it took a long time to build up a large planet. A calculation made by George W. Wetherill of the Carnegie Institution of Washington suggests that about 100 million years could pass between the formation of an object measuring 10 kilometers in diameter and an object the size of the earth.

The process of accretion had significant thermal consequences for the earth, consequences that have forcefully directed its evolution. Large bodies slamming into the planet produced immense heat in the interior, melting the cosmic dust found there. The resulting furnace—situated some 200 to 400 kilometers underground and called a magma ocean—

was active for millions of years, giving rise to volcanic eruptions. When the earth was young, heat at the surface caused by volcanism and lava flows from the interior was supplemented by the constant bombardment of huge planetesimals, some of them perhaps the size of the moon or even Mars. No life was possible during this period.

Beyond clarifying that the earth had formed through accretion, the Apollo program compelled scientists to try to reconstruct the subsequent temporal and physical development of the early earth. This undertaking had been considered impossible by founders of geology, including Charles Lyell, to whom the following phrase is attributed: No vestige of a beginning, no prospect for an end. This statement conveys the idea that the young earth could not be re-created, because its remnants were destroyed by its very activity. But the development of isotope geology in the 1960s had rendered this view obsolete. Their imaginations fired by Apollo and the moon findings, geochemists began to apply this technique to understand the evolution of the earth.

Figure 4.1 EARTH SEEN FROM SPACE has changed dramatically. One hundred million years after it had formed—some 4.35 billion years ago—the planet was probably undergoing meteor bombardment (*left*). At this time, it may have been studded with volcanic islands and shrouded by an atmosphere laden with carbon dioxide and heavy with clouds. Three billion years ago its face may have been obscured by an orange haze of methane, the product of early organisms (*center*). Today clouds, oceans and continents are clearly discernible (*right*). (This figure was prepared with the help of James F. Kasting of Pennsylvania State University.)

Dating rocks using so-called radioactive clocks allows geologists to work on old terrains that do not contain fossils. The hands of a radioactive clock are isotopes—atoms of the same element that have different atomic weights—and geologic time is measured by the rate of decay of one isotope into another [see "The Earliest History of the Earth," by Derek York; SCIENTIFIC AMERICAN, January 1993]. Among the many clocks, those based on the decay of uranium 238 into lead 206 and of uranium 235 into lead 207 are special. Geochronologists can determine the age of samples by analyzing only the daughter product—in this case, lead—of the radioactive parent, uranium.

Isotope geology has permitted geologists to determine that the accretion of the earth culminated in the differentiation of the planet: the creation of the core—the source of the earth's magnetic field—and the beginning of the atmosphere. In 1953 the classic work of Claire C. Patterson of the California Institute of Technology used the uranium-lead clock to establish an age of 4.55 billion

years for the earth and many of the meteorites that formed it. Recent work by one of us (Allègre) on lead isotopes, however, led to a somewhat new interpretation. As Patterson argued, some meteorites were indeed formed about 4.56 billion years ago, and their debris constituted the earth. But the earth continued to grow through the bombardment of planetesimals until some 120 to 150 million years later. At that time—4.44 to 4.41 billion years ago—the earth began to retain its atmosphere and create its core (see boxed figure "How the Earth Got Its Core"). This possibility had already been suggested by Bruce R. Doe and Robert E. Zartman of the U.S. Geological Survey in Denver a decade ago and is in agreement with Wetherill's estimates.

The emergence of the continents came somewhat later (see Figure 4.2). According to the theory of plate tectonics, these landmasses are the only part of the earth's crust that is not recycled and, consequently, destroyed during the geothermal cycle driven by the convection in the mantle. Continents thus provide a form of memory because the record of early life can be read in their rocks. The testimony, however, is not extensive. Geologic activity, including plate tectonics, erosion and metamorphism, has destroyed almost all the ancient rocks. Very few fragments have survived this geologic machine.

Nevertheless, in recent years, several important finds have been made, again using isotope geochemistry. One group, led by Stephen Moorbath of the University of Oxford, discovered terrain in West Greenland that is between 3.7 and 3.8 billion years old. In addition, Samuel A. Bowring of the Massachusetts Institute of Technology explored a small area in North America—the Acasta gneiss—that is 3.96 billion years old.

Ultimately, a quest for the mineral zircon led other researchers to even more ancient terrain. Typically found in continental rocks, zircon is not dissolved during the process of erosion but is deposited as particles in sediment. A few pieces of zircon can therefore survive for billions of years and can serve as a witness to the earth's more ancient crust. The search for old zircons started in Paris with the work of Annie Vitrac and Joël R. Lancelot, now at the University of Marseilles and the University of Montpellier, respectively, as well as with the efforts of Moorbath and Allègre. It was a group at the Australian National University in Canberra, directed by William Compston, that was finally successful. The team discovered zircons in western Australia that were between 4.1 and 4.3 billion years old.

Zircons have been crucial not only for understanding the age of the continents but for determining when life first appeared. The earliest fossils of undisputed age were found in Australia and South Africa. These relics of blue-green algae are about 3.5 billion years old. Manfred Schidlowski of the Max Planck Institute for Chemistry in Mainz has studied the Isua formation in West Greenland and argues that organic matter existed as many as 3.8 billion years ago. Because most of the record of

How the Earth Got Its Core

The differentiation of the planet took place quite quickly after the earth was formed by the accretion of cosmic dust and meteorites. About 4.4 billion years ago the core—which, with the mantle, drives the geothermal cycle, including volcanism—appeared; gases emerging from the interior of the planet also gave rise to a nascent atmosphere. Somewhat later, although the issue has not been entirely resolved, it seems that continental crust formed as the various elements segregated into different depths.

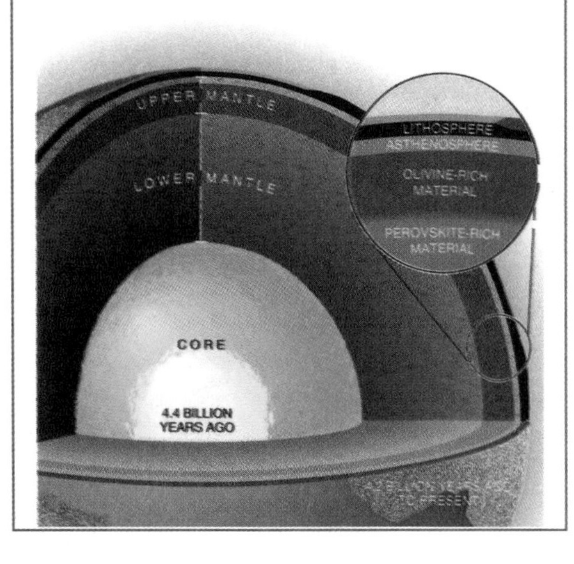

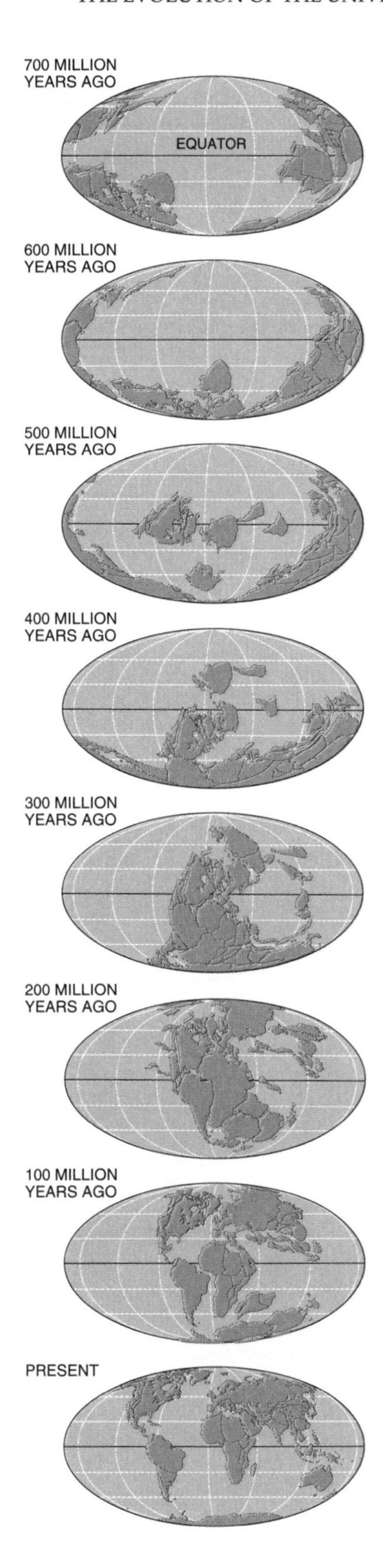

Figure 4.2 CONTINENTAL SHIFT has altered the face of the planet for nearly a billion years, as can be seen in the differences between the positions of the continents that we know today and those of 700 million years ago. Pangaea, the superaggregate of early continents, came together about 200 million years ago and then promptly, in geologic terms, broke apart. (This series was compiled with the advice of Christopher R. Scotese of the University of Texas at Arlington.)

early life has been destroyed by geologic activity, we cannot say exactly when it first appeared—perhaps it arose very quickly, maybe even 4.2 billion years ago.

One of the most important aspects of the earth's evolution is the formation of the atmosphere, because it is this assemblage of gases that allowed life to crawl out of the oceans and to be sustained. Researchers have hypothesized since the 1950s that the terrestrial atmosphere was created by gases emerging from the interior of the planet. When a volcano spews gases, it is an example of the continuous outgassing, as it is called, of the earth. But scientists have questioned whether this process occurred suddenly about 4.4 billion years ago when the core differentiated or whether it took place gradually over time.

To answer this question, Allègre and his colleagues studied the isotopes of rare gases. These gases—including helium, argon and xenon—have the peculiarity of being chemically inert, that is, they do not react in nature with other elements. Two of them are particularly important for atmospheric studies: argon and xenon. Argon has three isotopes, of which argon 40 is created by the decay of potassium 40. Xenon has nine, of which 129 has two different origins. Xenon 129 arose as the result of nucleosynthesis before the earth and solar system were formed. It was also created from the decay of radioactive iodine 129, which does not exist on the earth anymore. This form of iodine was present very early on but has died since, and xenon 129 has grown at its expense.

Like most couples, both argon 40 and potassium 40 and xenon 129 and iodine 129 have stories to tell. They are excellent chronometers. Although the atmosphere was formed by the outgassing of the mantle, it does not contain any potassium 40 or iodine 129. All argon 40 and xenon 129, formed in the earth and released, are found in the atmosphere today. Xenon was expelled from the mantle and retained in the atmosphere; therefore, the atmosphere-mantle ratio of this element allows us to evaluate the age of differentiation. Argon and xenon trapped in the mantle evolved by the radioactive decay of potassium 40 and iodine 129. Thus, if the total outgassing of the mantle occurred at the beginning of the earth's formation, the atmosphere would not contain any argon 40 but would contain xenon 129.

The major challenge facing an investigator who wants to measure such ratios of decay is to obtain high concentrations of rare gases in mantle rocks because they are extremely limited. Fortunately, a natural phenomenon occurs at mid-oceanic ridges during which volcanic lava transfers some silicates from the mantle to the surface. The small amounts of gases trapped in mantle minerals rise with the melt to the surface and are concentrated in small vesicles in the outer glassy margin of lava flows. This process serves to concentrate the amounts of mantle gases by a factor of 10^4 or 10^5. Collecting these rocks by dredging the seafloor and then crushing them under vacuum in a sensitive mass spectrometer allow geochemists to determine the ratios of the isotopes in the mantle. The results are quite surprising. Calculations of the ratios indicate that between 80 and 85 percent of the atmosphere was outgassed in the first one million years; the rest was released slowly but constantly during the next 4.4 billion years.

The composition of this primitive atmosphere was most certainly dominated by carbon dioxide, with nitrogen as the second most abundant gas (see Figure 4.3). Trace amounts of methane, ammonia, sulfur dioxide and hydrochloric acid were also present, but there was no oxygen. Except for the presence of abundant water, the atmosphere was similar to that of Venus or Mars. The details of the evolution of the original atmosphere are debated, particularly because we do not know how strong the sun was at that time. Some facts, however, are not disputed. It is evident that carbon dioxide played a crucial role. In addition, many scientists believe the evolving atmosphere contained sufficient quantities of gases like ammonia and methane to give rise to organic matter.

Still, the problem of the sun remains unresolved. One hypothesis holds that during the Archean era, which lasted from about 4.5 to 2.5 billion years ago, the sun's power was only 75 percent of what it is today. This possibility raises a dilemma: How could life have survived in the relatively cold climate that should accompany a weaker sun? A solution to the faint early sun paradox, as it is called, was offered by Carl Sagan and George Mullen of Cornell University in 1970. The two scientists suggested that methane and ammonia, which are very effective at trapping infrared radiation, were quite abundant. These gases could have created a super-greenhouse effect. The idea was criticized on the basis that such gases were highly reactive and have short lifetimes in the atmosphere.

In the late 1970s Veerabhadran Ramanathan, now at the Scripps Institution of Oceanography, and Robert D. Cess and Tobias Owen of the State University of New York at Stony Brook proposed another solution. They postulated that there was no need for methane in the early atmosphere because carbon dioxide was abundant enough to bring about the super-greenhouse effect. Again this argument raised a different question: How much carbon dioxide was there in the early atmosphere? Terrestrial carbon dioxide is now buried in carbonate rocks, such as limestone, although it is not clear when it became trapped there. Today calcium carbonate is created primarily during biological activity; in the Archean period, carbon may have been primarily removed during inorganic reactions.

The rapid outgassing of the planet liberated voluminous quantities of water from the mantle, creating the oceans and the hydrologic cycle. The acids that were probably present in the atmosphere eroded rocks, forming carbonate-rich rocks. The relative importance of such a mechanism is, however, debated. Heinrich D. Holland of Harvard University believes the amount of carbon dioxide in the atmosphere rapidly decreased during the Archean and stayed at a low level.

Understanding the carbon dioxide content of the early atmosphere is pivotal to understanding the mechanisms of climatic control. Two sometimes conflicting camps have put forth ideas on how this process works. The first group holds that global temperatures and carbon dioxide were controlled by inorganic geochemical feedbacks; the second asserts that they were controlled by biological removal.

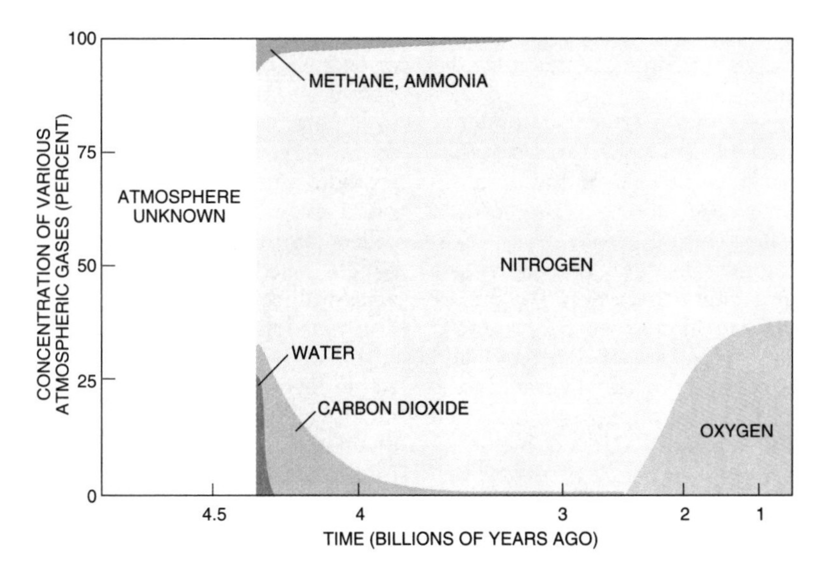

Figure 4.3 ATMOSPHERIC COMPOSITION, shown by the relative concentration of various gases, has been greatly influenced by life on the earth. The early atmosphere had fairly high concentrations of water and carbon dioxide and, some experts believe, methane, ammonia and nitrogen. After the emergence of living organisms, the oxygen that is so vital to our survival became more plentiful. Today carbon dioxide, methane and water exist only in trace amounts in the atmosphere.

James C. G. Walker, James F. Kasting and Paul B. Hays, then at the University of Michigan, proposed the inorganic model in 1981. They postulated that levels of the gas were high at the outset of the Archean and did not fall precipitously. The trio suggested that as the climate warmed, more water evaporated, and the hydrologic cycle became more vigorous, increasing precipitation and runoff. The carbon dioxide in the atmosphere mixed with rainwater to create carbonic acid runoff, exposing minerals at the surface to weathering. Silicate minerals combined with carbon that had been in the atmosphere, sequestering it in sedimentary rocks. Less carbon dioxide in the atmosphere meant, in turn, less of a greenhouse effect. The inorganic negative feedback process offset the increase in solar energy.

This solution contrasts with a second paradigm: biological removal. One theory advanced by James E. Lovelock, an originator of the Gaia hypothesis, assumed that photosynthesizing microorganisms, such as phytoplankton, would be very productive in a high-carbon dioxide environment. These creatures slowly removed carbon dioxide from the air and oceans, converting it into calcium carbonate sediments. Critics retorted that phytoplankton had not even evolved for most of the time that the earth has had life. (The Gaia hypothesis holds that life on the earth has the capacity to regulate temperature and the composition of the earth's surface and to keep it comfortable for living organisms.)

More recently, Tyler Volk of New York University and David W. Schwartzman of Howard University proposed another Gaian solution. They noted that bacteria increase carbon dioxide content in soils by breaking down organic matter and by generating humic acids. Both activities accelerate weathering, removing carbon dioxide from the atmosphere. On this point, however, the controversy becomes acute. Some geochemists, including Kasting, now at Pennsylvania State University, and Holland, postulate that while life may account for some carbon dioxide removal after the Archean, inorganic geochemical processes can explain most of the sequestering. These researchers view life as a rather weak climatic stabilizing mechanism for the bulk of geologic time.

The issue of carbon remains critical to the story of how life influenced the atmosphere. Carbon burial is a key to the vital process of building up atmospheric oxygen concentrations—a prerequisite for the development of certain life-forms. In addition, global warming may be taking place now as a result of humans releasing this carbon. For one or two billion years, algae in the oceans produced oxygen. But because this gas is highly reactive and because there were many reduced minerals in the ancient oceans—iron, for example, is easily oxidized—much of the oxygen produced by living creatures simply got used up before it could reach the atmosphere, where it would have encountered gases that would react with it.

Even if evolutionary processes had given rise to more complicated life-forms during this anaerobic era, they would have had no oxygen. Furthermore, unfiltered ultraviolet sunlight would have likely killed them if they left the ocean. Researchers such as Walker and Preston Cloud, then at the University of California at Santa Barbara, have suggested that only about two billion years ago, after most of the reduced minerals in the sea were oxidized, did atmospheric oxygen accumulate. Between one and two billion years ago oxygen reached current levels, creating a niche for evolving life.

By examining the stability of certain minerals, such as iron oxide or uranium oxide, Holland has shown that the oxygen content of the Archean atmosphere was low, before two billion years ago. It is largely agreed that the present-day oxygen content of 20 percent is the result of photosynthetic activity. Still, the question is whether the oxygen content in the atmosphere increased gradually over time or suddenly. Recent studies indicate that the increase of oxygen started abruptly between 2.1 and 2.03 billion years ago, and the present situation was reached 1.5 billion years ago.

The presence of oxygen in the atmosphere had another major benefit for an organism trying to live at or above the surface: it filtered ultraviolet radiation. Ultraviolet radiation breaks down many molecules—from DNA and oxygen to the chlorofluorocarbons that are implicated in stratospheric ozone depletion. Such energy splits oxygen into the highly unstable atomic form O, which can combine back into O_2 and into the very special molecule O_3, or ozone. Ozone, in turn, absorbs ultraviolet radiation. It was not until oxygen was abundant enough in the atmosphere to allow the formation of ozone that life even had a chance to get a root-hold or a foothold on land. It is not a coincidence that the rapid evolution of life from prokaryotes (single-celled organisms with no nucleus) to eukaryotes (single-celled organisms with a nucleus) to metazoa (multicelled organisms) took place in the billion-year-long era of oxygen and ozone.

Although the atmosphere was reaching a fairly stable level of oxygen during this period, the climate was hardly uniform. There were long stages of relative warmth or coolness during the transition to modern geologic time. The composition of fossil plankton shells that lived near the ocean floor provides a measure of bottom water temperatures. The record suggests that over the past 100 million years bottom waters cooled by nearly 15 degrees Celsius. Sea levels dropped by hundreds of meters, and continents drifted apart. Inland seas mostly disappeared, and the climate cooled an average of 10 to 15 degrees C. Roughly 20 million years ago permanent ice appears to have built up on Antarctica.

About two to three million years ago the paleoclimatic record starts to show significant expansions and contractions of warm and cold periods on 40,000-year or so cycles. This periodicity is interesting because it corresponds to the time it takes the earth to complete an oscillation of the tilt of its axis of rotation. It has long been speculated, and recently calculated, that known changes in orbital geometry could alter the amount of sunlight coming in between winter and summer by about 10 percent or so and could be responsible for initiating or ending ice ages.

Most interesting and perplexing is the discovery that between 600,000 and 800,000 years ago the dominant cycle switched from 40,000-year periods to 100,000-year intervals with very large fluctuations. The last major phase of glaciation ended about 10,000 years ago. At its height 20,000 years ago, ice sheets a mile thick covered much of northern Europe and North America. Glaciers expanded in high plateaus and mountains throughout the world. Enough ice was locked up on land to cause sea levels to drop more than 100 meters below where they are today. Massive ice sheets scoured the land and revamped the ecological face of the earth, which was five degrees C cooler on average than it is currently.

The precise causes of these changes are not yet sorted out. Volcanic eruptions may have played a significant role as shown by the effect of El Chichón in Mexico and Mount Pinatubo in the Philippines. Tectonic events, such as the development of the Himalayas, may influence world climate. Even the impact of comets can influence short-term climatic trends with catastrophic consequences for life [see "What Caused the Mass Extinction? An Extraterrestrial Impact," by Walter Alvarez and Frank Asaro; and "What Caused the Mass Extinction? A Volcanic Eruption," by Vincent E. Courtillot; SCIENTIFIC AMERICAN, October 1990]. It is remarkable that despite violent, episodic perturbations, the climate has been buffered enough to sustain life for 3.5 billion years.

One of the most pivotal climatic discoveries of the past 20 years has come from ice cores in Greenland and Antarctica (see Figure 4.4). When snow falls on these frozen continents, the air between the snow grains is trapped as bubbles. The snow is gradually compressed into ice, along with its captured gases. Some of these records can go back as far as 200,000 years; scientists can analyze the chemical content of ice and bubbles from sections of ice that lie as deep as 2,000 meters below the surface.

The ice-core borers have determined that the air breathed by ancient Egyptians and Anasazi Indians was very similar to that which we inhale today—except for a host of air pollutants introduced over the past 100 or 200 years. Principal among these added gases, or pollutants, are extra carbon dioxide and methane. The former has increased 25 percent as a result of industrialization and deforestation; the latter has doubled because of agriculture, land use and energy production. The concern that increased amounts of these gases might trap enough heat to cause global warming is at the heart of the climate debate [see "The Changing Climate," by Stephen H. Schneider; SCIENTIFIC AMERICAN, September 1989].

The ice cores have shown that sustained natural rates of worldwide temperature change are typically about one degree C per millennium. These shifts are still significant enough to have radically altered where species live and to have potentially contributed to the extinction of such charismatic megafauna as mammoths and saber-toothed tigers. But a most extraordinary story from the ice cores is not the relative stability of the climate during the past 10,000 years. It appears that during the height of the last ice age 20,000 years ago there was between 30 and 40 percent less carbon dioxide and 50 percent less methane in the air than there has been during our period, the Holocene. This finding suggests a positive feedback between carbon dioxide, methane and climatic change.

The reasoning that supports the idea of this destabilizing feedback system goes as follows. When the world was colder, there was less concentration of greenhouse gases, and so less heat was trapped. As the earth warmed up, carbon dioxide and methane levels increased, accelerating the warming. If life had a hand in this story, it would have been to drive, rather than to oppose, climatic change (see Figure 4.5). Once again, though, this picture is incomplete.

Figure 4.4 ICE CORES from Greenland or Antarctica have provided scientists with a swatch cut from the earth's atmospheric history. As snow is compressed into ice, air bubbles trapped between the flakes are preserved. By analyzing the gases in these tiny chambers, researchers can determine the composition of the atmosphere up to almost 200,000 years ago.

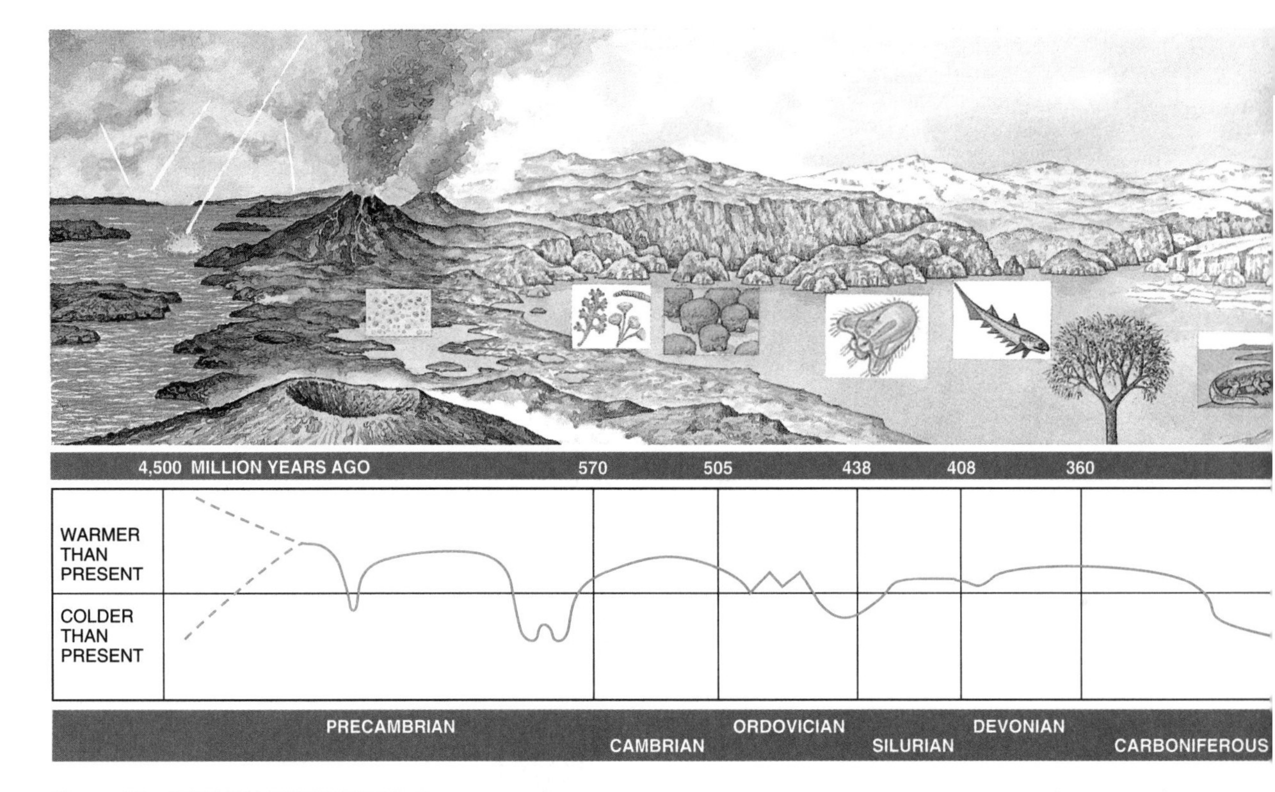

Figure 4.5 CLIMATE FLUCTUATIONS are apparent over time. Although the earth's early temperature record is quite uncertain, good estimates can be made starting 400 million years ago, when fossils were more abundantly preserved. As climate shifted, so did life—suggesting feedback between the two. The dates of these evolutions remain unclear as well,

Nevertheless, most scientists would agree that life could well be the principal factor in the positive feedback between climatic change and greenhouse gases. One hypothesis suggests that increased nutrient runoff from the continental shelves that were exposed as sea levels fell fertilized phytoplankton. This nutrient input could have created a larger biomass of such marine species. Because the calcium carbonate shell makes up most of the mass of many phytoplankton species, increased productivity would remove carbon dioxide from the oceans and eventually the atmosphere. At the same time, boreal forests that account for about 10 to 20 percent of the carbon in the atmosphere were decimated during the ice ages. Carbon from these high-latitude forests could have been released to the atmosphere, yet the atmosphere had less of this gas then. Thus, the positive feedback system of the ocean's biological pump may have offset the negative feedback caused by the destruction of the forests. Great amounts of carbon can be stored in soils, however, so the demise of forests may have led to the sequestering of carbon in the ground.

What is significant is the idea that the feedback was positive. By studying the transition from the high carbon dioxide–low oxygen atmosphere of the Archean to the era of great evolutionary progress about a half a billion years ago, it becomes clear that life may have been a factor in the stabilization of climate. In another example—during the ice ages and interglacial cycles—life seems to have the opposite function: accelerating the change rather than diminishing it. This observation has led one of

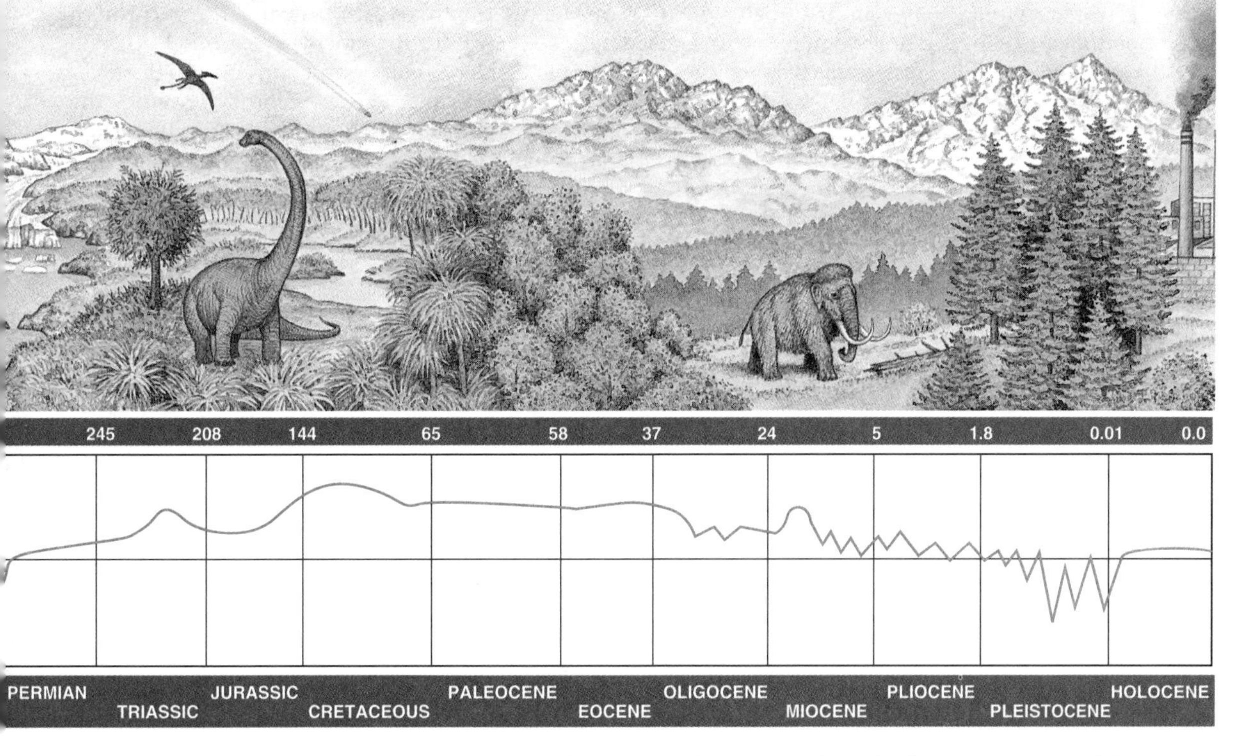

but their order is more apparent. First a primordial soup formed, then primitive organisms, such as algae, stromatolites and jellyfish, arose; spiny fish were followed by the ichthyostega, perhaps the first creature to crawl from ocean onto land. The rest of the story is well known: dinosaurs appeared and died out, their place taken by mammals.

us (Schneider) to contend that climate and life coevolved rather than life serving solely as a negative feedback on climate.

If we humans consider ourselves part of life—that is, part of the natural system—then it could be argued that our collective impact on the earth means we may have a significant coevolutionary role in the future of the planet. The current trends of population growth, the demands for increased standards of living and the use of technology and organizations to attain these growth-oriented goals all contribute to pollution. When the price of polluting is low and the atmosphere is used as a free sewer, carbon dioxide, methane, chlorofluorocarbons, nitrous oxides, sulfur oxides and other toxics can build up.

The theory of heat trapping—codified in mathematical models of the climate—suggests that if carbon dioxide levels double sometime in the middle of the next century, the world will warm between one and five degrees C. The mild end of that range entails warming at the rate of one degree per 100 years—a factor of 10 faster than the one degree per 1,000 years that has historically been the average rate of natural change on a global scale. Should the higher end of the range occur, then we could see rates of climatic change 50 times faster than natural average conditions. Change at this rate would almost certainly force many species to attempt to move their ranges, just as they did from the ice age–interglacial transition between 10,000 and 15,000 years ago. Not only would species have to respond to climatic change at rates 10 to 50 times faster, but few would have undisturbed, open migration routes as they did at the end of the ice age and the onset of the interglacial era. It is for these reasons that it is essential to understand whether doubling carbon dioxide will warm the earth by one degree or five.

To make the critical projections of future climatic change needed to understand the fate of ecosystems on this earth, we must dig through land, sea and ice to uncover as much of the geologic, paleoclimatic and paleoecological records as we can. These records provide the backdrop against which to calibrate the crude instruments we must use to peer into a shadowy environmental future, a future increasingly influenced by us.

The Origin of Life on the Earth

Growing evidence supports the idea that the emergence of catalytic RNA was a crucial early step. How that RNA came into being remains unknown.

• • •

Leslie E. Orgel

When the earth formed some 4.6 billion years ago, it was a lifeless, inhospitable place. A billion years later it was teeming with organisms resembling blue-green algae. How did they get there? How, in short, did life begin? This long-standing question continues to generate fascinating conjectures and ingenious experiments, many of which center on the possibility that the advent of self-replicating RNA was a critical milestone on the road to life (see Figure 5.1).

Before the mid-17th century, most people believed that God had created humankind and other higher organisms and that insects, frogs and other small creatures could arise spontaneously in mud or decaying matter. For the next two centuries, those ideas were subjected to increasingly severe criticism, and in the mid-19th century two important scientific advances set the stage for modern discussions of the origin of life.

In one advance Louis Pasteur discredited the concept of spontaneous generation. He offered proof that even bacteria and other microorganisms arise from parents resembling themselves. He thereby highlighted an intriguing question: How did the first generation of each species come into existence?

The second advance, the theory of natural selection, suggested an answer. According to this proposal, set forth by Charles Darwin and Alfred Russel Wallace, some of the differences between individuals in a population are heritable. When the environment changes, individuals bearing traits that provide the best adaptation to the new environment meet with the greatest reproductive success. Consequently, the next generation contains an increased percentage of well-adapted individuals displaying the helpful characteristics. In other words, environmental pressures select adaptive traits for perpetuation.

Repeated generation after generation, natural selection could thus lead to the evolution of complex organisms from simple ones. The theory therefore implied that all current life-forms could have evolved from a single, simple progenitor—an organism now referred to as life's last common ancestor. (This life-form is said to be "last" not "first" because it is the nearest shared ancestor of all contemporary organisms; more distant ancestors must have appeared earlier.)

Darwin, bending somewhat to the religious biases of his time, posited in the final paragraph of *The Origin of Species* that "the Creator" originally

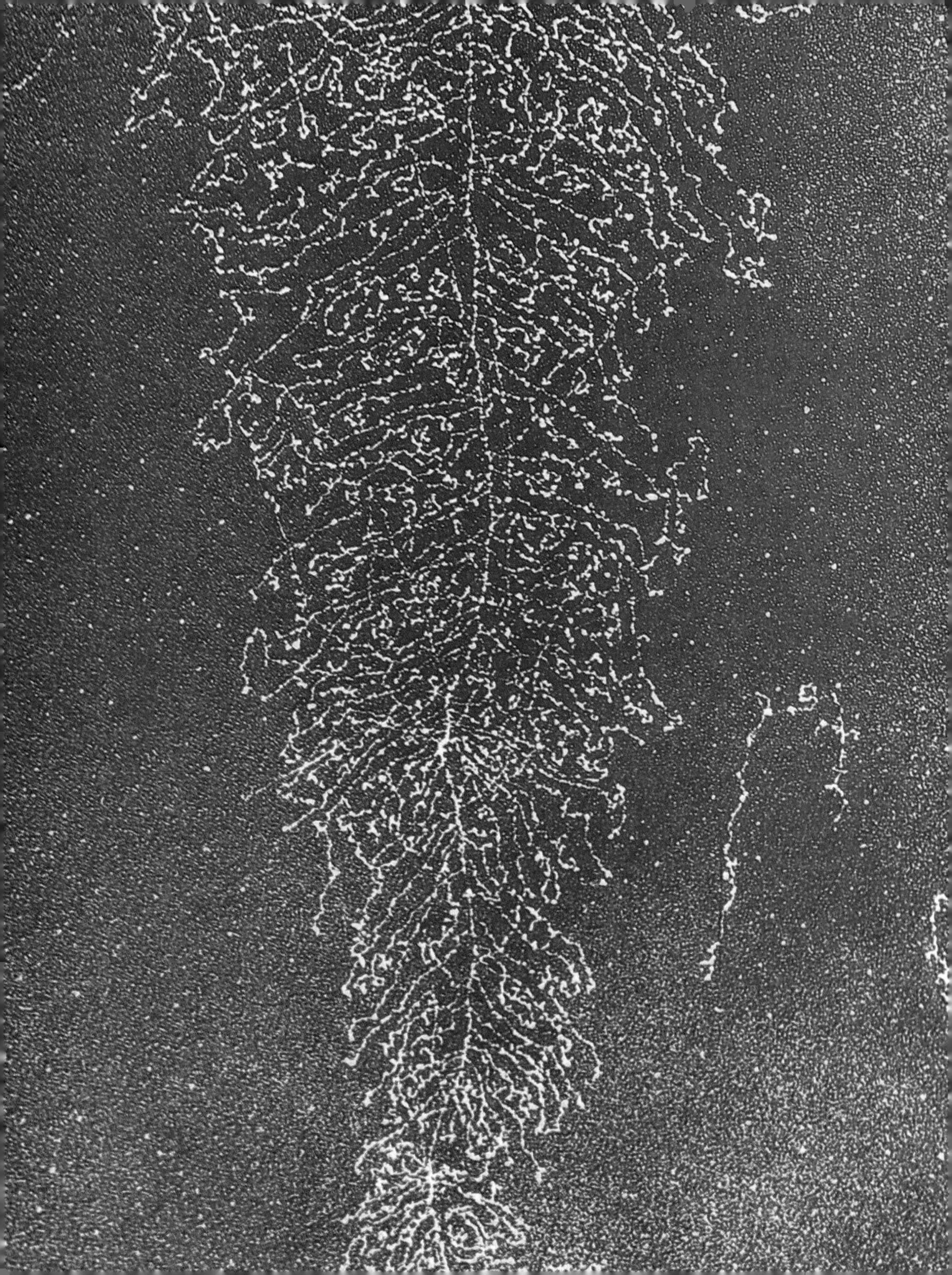

breathed life "into a few forms or into one." Then evolution took over: From so simple a beginning endless forms most beautiful and most wonderful have been, and are being evolved." In private correspondence, however, he suggested life could have arisen through chemistry, "in some warm little pond, with all sorts of ammonia and phosphoric salts, light, heat, electricity, etc. present." For much of the 20th century, origin-of-life research has aimed to flesh out Darwin's private hypothesis—to elucidate how, without supernatural intervention, spontaneous interaction of the relatively simple molecules dissolved in the lakes or oceans of the prebiotic world could have yielded life's last common ancestor.

Finding a solution to this problem requires knowing something about that ancestor's characteristics. Obviously, it had to possess genetic information—that is, heritable instructions for functioning and reproducing—and the means to replicate and carry out those instructions. Otherwise it would have left no descendants. Also, its system for replicating its genetic material had to allow for some random variation in the heritable characteristics of the offspring so that new traits could be selected and lead to the creation of diverse species.

Scientists have attained more insight into the character of the last common ancestor by identifying commonalities in contemporary organisms. One can safely infer that intricate features present in all modern varieties of life also appeared in that common ancestor. After all, it is next to impossible for such universal traits to have evolved separately. The rationale is the same as would apply to discovery of two virtually identical screenplays, differing only in a few words. It would be unreasonable to think that the scripts were created independently by two separate authors. By the same token, it would be safe to assume that one script was an imperfect replica of the other or that both versions were slightly altered copies of a third.

Figure 5.1 STRANDS OF RNA (*branchlike lines*) are being synthesized on DNA (*vertical line*). Today genetic information usually flows from DNA to RNA, but many investigators think some form of RNA evolved before DNA. This idea is central to the RNA-world theory of how life began, which holds that RNA made possible the evolution of DNA and life itself.

One readily apparent commonality is that all living things consist of similar organic (carbon-rich) compounds. Another shared property is that the proteins found in present-day organisms are fashioned from one set of 20 standard amino acids. These proteins include enzymes (biological catalysts) that are essential to development, survival and reproduction.

Further, contemporary organisms carry their genetic information in nucleic acids—RNA and DNA—and use essentially the same genetic code. This code specifies the amino acid sequences of all the proteins each organism needs. More precisely, the instructions take the form of specific sequences of nucleotides, the building blocks of nucleic acids. These nucleotides consist of a sugar (deoxyribose in DNA, and ribose in RNA), a phosphate group and one of four different nitrogen-containing bases. In DNA, the bases are adenine (*A*), guanine (*G*), cytosine (*C*) and thymine (*T*). In RNA, uracil (*U*) substitutes for thymine (see Figure 5.2). The bases constitute the alphabet, and triplets of bases form the words. As an example, the triplet *CUU* in RNA instructs a cell to add the amino acid leucine to a growing strand of protein.

From such findings we can infer that our last common ancestor stored genetic information in nucleic acids that specified the composition of all needed proteins. It also relied on proteins to direct many of the reactions required for self-perpetuation. Hence, the central problem of origin-of-life research can be refined to ask, By what series of chemical reactions did this interdependent system of nucleic acids and proteins come into being?

Anyone trying to solve this puzzle immediately encounters a paradox. Nowadays nucleic acids are synthesized only with the help of proteins, and proteins are synthesized only if their corresponding nucleotide sequence is present. It is extremely improbable that proteins and nucleic acids, both of which are structurally complex, arose spontaneously in the same place at the same time. Yet it also seems impossible to have one without the other. And so, at first glance, one might have to conclude that life could never, in fact, have originated by chemical means.

In the late 1960s Carl R. Woese of the University of Illinois, Francis Crick, then at the Medical Research Council in England, and I (working at the Salk Institute for Biological Studies in San Diego) independently suggested a way out of this difficulty. We proposed that RNA might well have

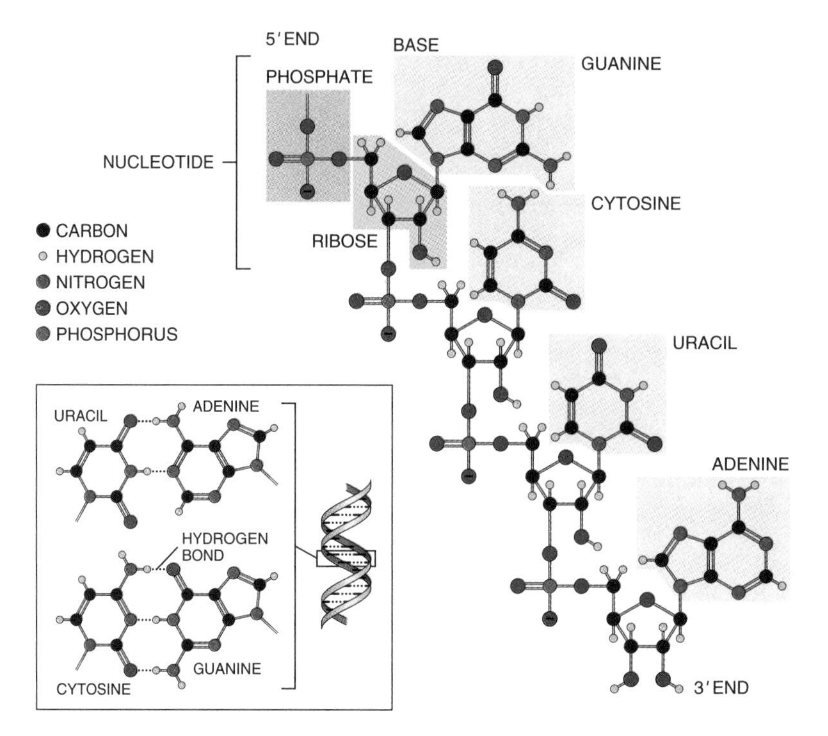

Figure 5.2 RNA IS COMPOSED OF nucleotides, each of which consists of a phosphate group (*green box*) and the sugar ribose (*yellow box*) linked to a nitrogen-containing base (*blue boxes*): guanine (*G*), cytosine (*C*), uracil (*U*) or adenine (*A*). Uracil on one strand of RNA can pair with adenine on another strand, and cytosine can pair with guanine, producing a double helix (*inset*). Such complementary base pairing is thought to have contributed to the ability of early RNA to reproduce itself.

come first and established what is now called the RNA world—a world in which RNA catalyzed all the reactions necessary for a precursor of life's last common ancestor to survive and replicate. We also posited that RNA could subsequently have developed the ability to link amino acids together into proteins. This scenario could have occurred, we noted, if prebiotic RNA had two properties not evident today: a capacity to replicate without the help of proteins and an ability to catalyze every step of protein synthesis.

There were a few reasons why we favored RNA over DNA as the originator of the genetic system, even though DNA is now the main repository of hereditary information. One consideration was that the ribonucleotides in RNA are more readily synthesized than are the deoxyribonucleotides in DNA. Moreover, it was easy to envision ways that DNA could evolve from RNA and then, being more stable, take over RNA's role as the guardian of heredity. We suspected that RNA came before proteins in part because we had difficulty composing any scenario in which proteins could replicate in the absence of nucleic acids.

During the past 10 years, a fair amount of evidence has lent credence to the idea that the hypothetical RNA world did exist and lead to the advent of life based on DNA, RNA and protein. Notably, in 1983 Thomas R. Cech of the University of Colorado at Boulder and, independently, Sidney Altman of Yale University discovered the first known ribozymes, enzymes made of RNA. Until then, proteins were thought to carry out all catalytic reactions in contemporary organisms. Indeed, the term "enzyme" is usually reserved for proteins. The first ribozymes identified could do little more than cut and join preexisting RNA. Nevertheless, the fact that they behaved like enzymes added weight to the notion that ancient RNA might also have been catalytic.

So far no RNA molecules that direct the replication of other RNA molecules have been identified in nature. But ingenious techniques devised by Cech and Jack W. Szostak of the Massachusetts General Hospital have modified naturally occurring ribozymes so that they can carry out some of the most important subreactions of RNA replication, such as stringing together nucleotides or oligonucleotides (short sequences of nucleotides).

Quite recently Szostak found even stronger evidence that an RNA molecule produced by prebiotic chemistry could have carried out RNA replication on the early earth. He started by creating a pool of random oligonucleotides, to approximate the random production presumed to have occurred some four billion years ago. From that pool he was able to isolate a catalyst that could join together oligonucleotides. Equally important, the catalyst could draw energy for the reaction from a triphosphate group (three joined phosphates), the very same group that now fuels most biochemical reactions in living systems, including nucleic acid replication. Such a resemblance supports the idea that an RNA molecule could have behaved like, and preceded, the protein catalysts that today carry out the replication of genetic material in living organisms. Much remains to be done, but it now seems likely that some kind of RNA-catalyzed reproduction of RNA will be demonstrated in the not too distant future.

Studies of ribosomes, often called the protein factories of cells, have provided support for another important part of the RNA-world hypothesis: the proposition that RNA could have created protein synthesis. Ribosomes, which consist of ribosomal RNA and protein, travel along strands of messenger RNA (single-strand transcripts of protein-coding genes carried by DNA). As the ribosomes move, they link one specified amino acid to the next by forming peptide bonds between them. Harry F. Noller, Jr., of the University of California at Santa Cruz has found that it is probably the RNA in ribosomes, not the protein, that catalyzes formation of the peptide bonds.

Other work indicates that primitive RNA would have been able to evolve, as would be required of any material that gave rise to the genes in life's last common ancestor. Sol Spiegelman, when at the University of Illinois, and researchers inspired by his ideas have demonstrated that RNA molecules can be induced to take on new traits. For instance, when RNA was allowed to replicate repeatedly in the presence of a ribonuclease (an enzyme that normally breaks down RNA), the RNA eventually became resistant to the degradative enzyme. Similarly, Gerald F. Joyce of the Scripps Research Institute and others have recently applied more sophisticated procedures to derive ribozymes that cleave a variety of chemical bonds, including peptide bonds.

Thus, there is good reason to think the RNA world did exist and that RNA invented protein synthesis. If this conclusion is correct, the main task of origin-of-life research then becomes explaining how the RNA world came into being. The answer to this question requires knowing something about the chemistry of the prebiotic soup: the aqueous solution of organic molecules in which life originated. Fortunately, even before the RNA-world hypothesis was proposed, investigators had gained useful insights into that chemistry.

By the 1930s Alexander I. Oparin in Russia and J. B. S. Haldane in England had pointed out that the organic compounds needed for life could not have formed on the earth if the atmosphere was as rich in oxygen (oxidizing) as it is today. Oxygen, which takes hydrogen atoms from other compounds, interferes with the reactions that transform simple organic molecules into complex ones. Oparin and Haldane proposed, therefore, that the atmosphere of the young earth, like that of the outer planets, was reducing: it contained very little oxygen and was rich in hydrogen (H_2) and compounds that can donate hydrogen atoms to other substances. Such gases were presumed to include methane (CH_4) and ammonia (NH_3).

Oparin's and Haldane's ideas inspired the famous Miller-Urey experiment, which in 1953 began the era of experimental prebiotic chemistry (see boxed figure "The Original Origin-of-Life Experiment"). Harold C. Urey of the University of

Chicago and Stanley L. Miller, a graduate student in Urey's laboratory, wondered about the kinds of reactions that occurred when the earth was still enveloped in a reducing atmosphere. In a self-contained apparatus, Miller created such an "atmosphere." It consisted of methane, ammonia, water and hydrogen above an "ocean" of water. Then he subjected the gases to "lightning" in the form of a continuous electrical discharge. After a few days, he analyzed the contents of the mock ocean.

Miller found that as much as 10 percent of the carbon in the system was converted to a relatively small number of identifiable organic compounds, and up to 2 percent of the carbon went to making amino acids of the kinds that serve as constituents of proteins. This last discovery was particularly exciting because it suggested that the amino acids needed for the construction of proteins—and for life itself—would have been abundant on the primitive planet. At the time, investigators were not yet paying much attention to the origin of nucleic acids; they were most interested in explaining how proteins appeared on the earth.

Careful analyses elucidated many of the chemical reactions that occurred in the experiment and thus might have occurred on the prebiotic planet. First, the gases in the "atmosphere" reacted to form a suite of simple organic compounds, including hydrogen cyanide (HCN) and aldehydes (compounds containing the group CHO). The aldehydes then combined with ammonia and hydrogen cyanide to generate intermediary products called aminonitriles, which interacted with water in the "ocean" to produce amino acids and ammonia. Glycine was the most abundant amino acid, resulting from the combination of formaldehyde (CH_2O), ammonia and hydrogen cyanide. A surprising number of the standard 20 amino acids were also made in lesser amounts.

Since then, workers have subjected many different mixtures of simple gases to various energy sources. The results of these experiments can be

The Original Origin-of-Life Experiment

In the early 1950s Stanley L. Miller, working in the laboratory of Harold C. Urey at the University of Chicago, did the first experiment designed to clarify the chemical reactions that occurred on the primitive earth (*right*). In the flask at the bottom, he created an "ocean" of water, which he heated, forcing water vapor to circulate (*arrows*) through the apparatus. The flask at the top contained an "atmosphere" consisting of methane (CH_4), ammonia (NH_3), hydrogen (H_2) and the circulating water vapor. Next he exposed the gases to a continuous electrical discharge ("lightning"), causing the gases to interact. Water-soluble products of those reactions then passed through a condenser and dissolved in the mock ocean. The experiment yielded many amino acids and enabled Miller to explain how they had formed. For instance, glycine appeared after reactions in the atmosphere produced simple compounds—formaldehyde and hydrogen cyanide—that participated in the set of reactions shown. Years after this experiment, a meteorite that struck near Murchison, Australia, was shown to contain a number of the same amino acids that Miller identified (*table*) and in roughly the same relative amounts (*dots*); those found in proteins are highlighted in blue. Such coincidences lent credence to the idea that Miller's protocol approximated the chemistry of the prebiotic earth. More recent findings have cast some doubt on that conclusion.

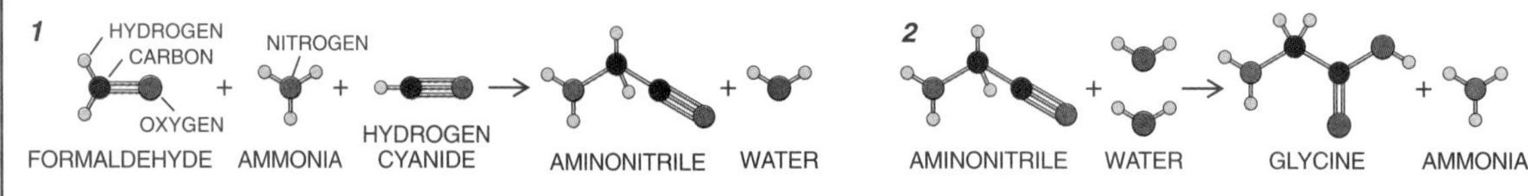

summarized neatly. Under sufficiently reducing conditions, amino acids form easily. Conversely, under oxidizing conditions, they do not arise at all or do so only in small amounts.

Similar studies provided some of the first evidence that the components of nucleic acids could have formed in the prebiotic soup as well. In 1961 Juan Oró, then at the University of Houston, tried to determine whether amino acids could be obtained by even simpler chemistry than had operated in the Miller-Urey experiment. He mixed hydrogen cyanide and ammonia in an aqueous solution, without introducing an aldehyde. He found that amino acids could indeed be produced from these chemicals. In addition, he made an unexpected discovery: the most abundant complex molecule identified was adenine.

Adenine, it will be recalled, is one of the four nitrogen-containing bases present in RNA and DNA. It is also a component of adenosine triphosphate (ATP), now the major energy-providing molecule of biochemistry. Oró's work implied that if the atmosphere was indeed reducing, adenine—arguably one of the most essential biochemicals—would have been available to help get life started. Later studies established that the remaining nucleic acid bases could be obtained from reactions among hydrogen cyanide and two other compounds that would have formed in a reducing prebiotic atmosphere: cyanogen (C_2N_2) and cyanoacetylene (HC_3N). Hence, early experiments seemed to indicate that under plausible prebiotic conditions, important constituents of proteins and nucleic acids could have been present on the early earth.

Strikingly, many of the same compounds generated in these various experiments have also been shown to exist in outer space. A family of amino acids that overlaps strongly with those formed in the Miller-Urey experiment has been identified in carbonaceous meteorites, along with the purine bases (adenine and guanine). Furthermore, the

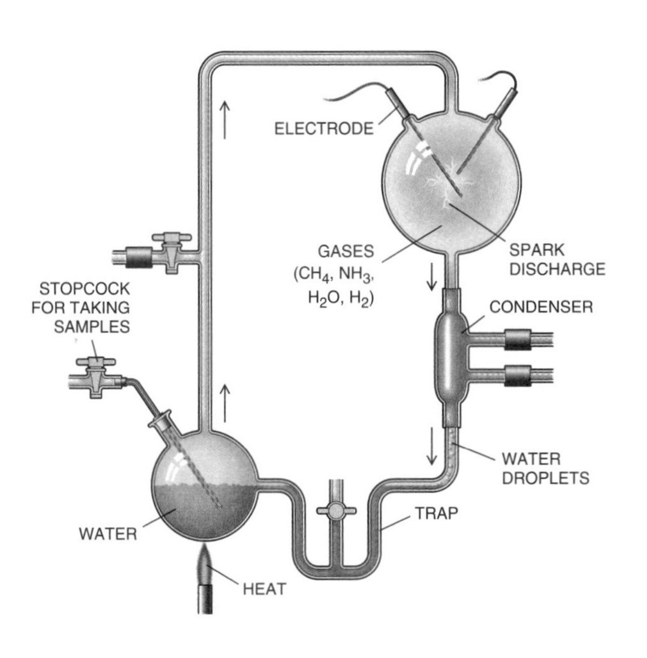

AMINO ACID	MURCHISON METEORITE	DISCHARGE EXPERIMENT
GLYCINE	• • • •	• • • •
ALANINE	• • • •	• • • •
α-AMINO-*N*-BUTYRIC ACID	• • •	• • • •
α-AMINOISOBUTYRIC ACID	• • • •	• •
VALINE	• • •	• •
NORVALINE	• • •	• • •
ISOVALINE	• •	• •
PROLINE	• • •	•
PIPECOLIC ACID	•	•
ASPARTIC ACID	• • •	• • •
GLUTAMIC ACID	• • •	• •
β-ALANINE	• •	• •
β-AMINO-*N*-BUTYRIC ACID	•	•
β-AMINOISOBUTYRIC ACID	•	•
γ-AMINOBUTYRIC ACID	•	• •
SARCOSINE	• •	• • •
N-ETHYLGLYCINE	• •	• • •
N-METHYLALANINE	• •	• •

family of small molecules that laboratory experiments have implicated as participating in prebiotic syntheses—water, ammonia, formaldehyde, hydrogen cyanide and cyanoacetylene—is abundant in interstellar dust clouds, where new stars are born.

The coincidence between the molecules present in outer space and those produced in laboratory simulations of prebiotic chemistry has generally been interpreted to mean that the simulations have painted a reasonable picture of the chemistry that occurred on the young earth. I should note, however, that this conclusion is now shakier than it once seemed. Doubt has arisen because recent investigations indicate the earth's atmosphere was never as reducing as Urey and Miller presumed. I suspect that many organic compounds generated in past studies would have been produced even in an atmosphere containing less hydrogen, methane and ammonia. Still, it seems prudent to consider other mechanisms for the accumulation of the constituents of proteins and nucleic acids in the prebiotic soup.

For instance, the amino acids and nitrogen-containing bases needed for life on the earth might have been delivered by interstellar dust, meteorites and comets. During the first half a billion years of the earth's history, bombardment by meteorites and comets must have been intense, although the extent to which organic material could have survived such impacts is debatable. It is also possible, though less likely, that some of the organic materials required for life did not originate at the earth's surface at all. They may have arisen in deep-sea vents, the submarine fissures in the earth's crust through which intensely hot gases are cycled.

Even if we assume that one process or another allowed the constituents of nucleic acids to appear on the prebiotic planet, those of us who favor the RNA-world hypothesis still have to explain how self-replicating RNA was created from these constituents. The simplest hypothesis presumes that the nucleotides in RNA formed when direct chemical reactions led to joining of the sugar ribose with nucleic acid bases and phosphate (which would have been available in inorganic material). Next, these ribonucleotides spontaneously joined to form polymers, at least one of which happened to be capable of engineering its own reproduction.

This scenario is attractive but, as will be seen, has proved hard to confirm. First of all, in the absence of enzymes, workers have had trouble synthesizing ribose in adequate quantity and purity. It has long been known that ribose can be produced easily through a series of reactions between molecules of formaldehyde. Yet when such reactions occur, they yield a mixture of sugars in which ribose is always a minor product. The relative paucity of ribose would militate against development of an RNA world, because the other sugars would combine with nucleic acid bases to form products that inhibit RNA synthesis and replication. No one has yet discovered a simple, complete chain of reactions that ends with ribose as the main product.

What is more, attempts to synthesize nucleotides directly from their components under prebiotic conditions have met with only modest success. One encouraging series of experiments has yielded purine nucleosides—that is, units consisting of ribose and a purine base but not including the phosphate group that would be present in a finished nucleotide. Unfortunately, investigators have been unable to produce pyrimidine nucleosides (combinations of ribose with cytosine or uracil) efficiently without the aid of enzymes.

Formation of nucleotides by combining phosphate with nucleosides has been achieved by simple prebiotic reactions. But the kinds of nucleotides that occur in nature arose along with related molecules having incorrect structures. If such mixtures were produced on the young planet, the abnormal nucleotides would have interacted with the normal ones to interfere with catalysis and RNA replication. Hence, although each step of ribonucleotide synthesis can be achieved to some extent, it is not easy to see how prebiotic reactions could have led to the development of the ribonucleotides needed for producing self-replicating RNA.

One way around this problem is to assume that inorganic catalysts were available to ensure that only the correct nucleotides formed. For instance, when the components of nucleotides became adsorbed on the surface of some mineral, that mineral might have caused them to combine only in specific orientations. The possibility that minerals served as useful catalysts remains real, but none of the minerals tested so far has been shown to have the specificity needed to yield only nucleotides having the correct architecture.

It is also possible that nonenzymatic reactions leading to the efficient synthesis of pure ribonucleotides did occur but that scientists have simply failed to identify them. As a case in point, Albert Eschenmoser of the Swiss Federal Institute of Technology recently managed to limit the number of different sugars generated when ribose was made from the polymerization of formaldehyde molecules. In his experiments, he substituted a normal intermediate of the ribose-forming reaction with a closely related, phosphorylated compound and then allowed the later steps to proceed. Under some conditions, the main end product of the process was a phosphorylated derivative of ribose. The phosphate groups on this product would have had to be rearranged in order to produce the phosphorylated ribose found in ribonucleotides. Nevertheless, the results do suggest that undiscovered reactions in the prebiotic soup could have led to the efficient synthesis of ribonucleotides.

Let us assume investigators could prove that ribonucleotides were able to emerge nonenzymatically. Workers who favor the simple scenario described above would still have to demonstrate that the nucleotides could assemble into polymers and that the polymers could replicate without assistance from proteins. Many researchers are now struggling with these challenges. Once again, minerals could conceivably have catalyzed the joining of reactive nucleotides into polymers. Indeed, James P. Ferris of the Rensselaer Polytechnic Institute finds that a common clay, montmorillonite, catalyzes the synthesis of RNA oligonucleotides.

It is harder to conceive of the steps by which RNA might have begun to replicate in the absence of proteins. Early work in my laboratory initially suggested that such replication was possible. In these experiments, we synthesized oligonucleotides and mixed them with free nucleotides. The nucleotides lined up on the oligonucleotides and combined with one another to form new oligonucleotides.

To be more specific, since 1953, when James D. Watson and Francis Crick solved the three-dimensional structure of DNA, it has been known that adenine in nucleotides pairs with thymine in DNA and with uracil in RNA. Similarly, guanine pairs with cytosine. Such coupled units are now known as Watson-Crick base pairs. The oligonucleotides that emerged in our experiments arose through Watson-Crick base pairing and were thus complementary to the original strands. For example, a template that was made solely of cytosine-bearing ribonucleotides directed construction of a complementary polymer consisting entirely of guanine-bearing ribonucleotides.

Forming such complements from an original template—a process I shall refer to as "copying"—would be the first step in prebiotic replication of a selected strand of RNA. Then the strands would have to separate, and a complement of the complement (a replica of the original strand) would have to be constructed. The experiments described above clearly established that the mutual attraction between adenine and uracil and between guanine and cytosine is sufficient by itself to yield complementary strands of many nucleotide sequences. Enzymes simply make the process more efficient and allow a broader range of RNAs to be copied.

After years of trying, however, we have been unable to achieve the second step of replication—copying of a complementary strand to yield a duplicate of the first template—without help from protein enzymes. Equally disappointing, we can induce copying of the original template only when we run our experiments with nucleotides having a right-handed configuration. All nucleotides synthesized biologically today are right-handed. Yet on the primitive earth, equal numbers of right- and left-handed nucleotides would have been present. When we put equal numbers of both kinds of nucleotides in our reaction mixtures, copying was inhibited.

All these problems are worrisome, but they do not completely rule out the possibility that RNA was initially synthesized and replicated by relatively uncomplicated processes. Perhaps minerals did indeed catalyze both the synthesis of properly structured nucleotides and their polymerization to a random family of oligonucleotides. Then copying without replication would have produced a pair of complementary strands. If, as Szostak has posited, one of the strands happened to be a ribozyme that could copy its complement and thus duplicate

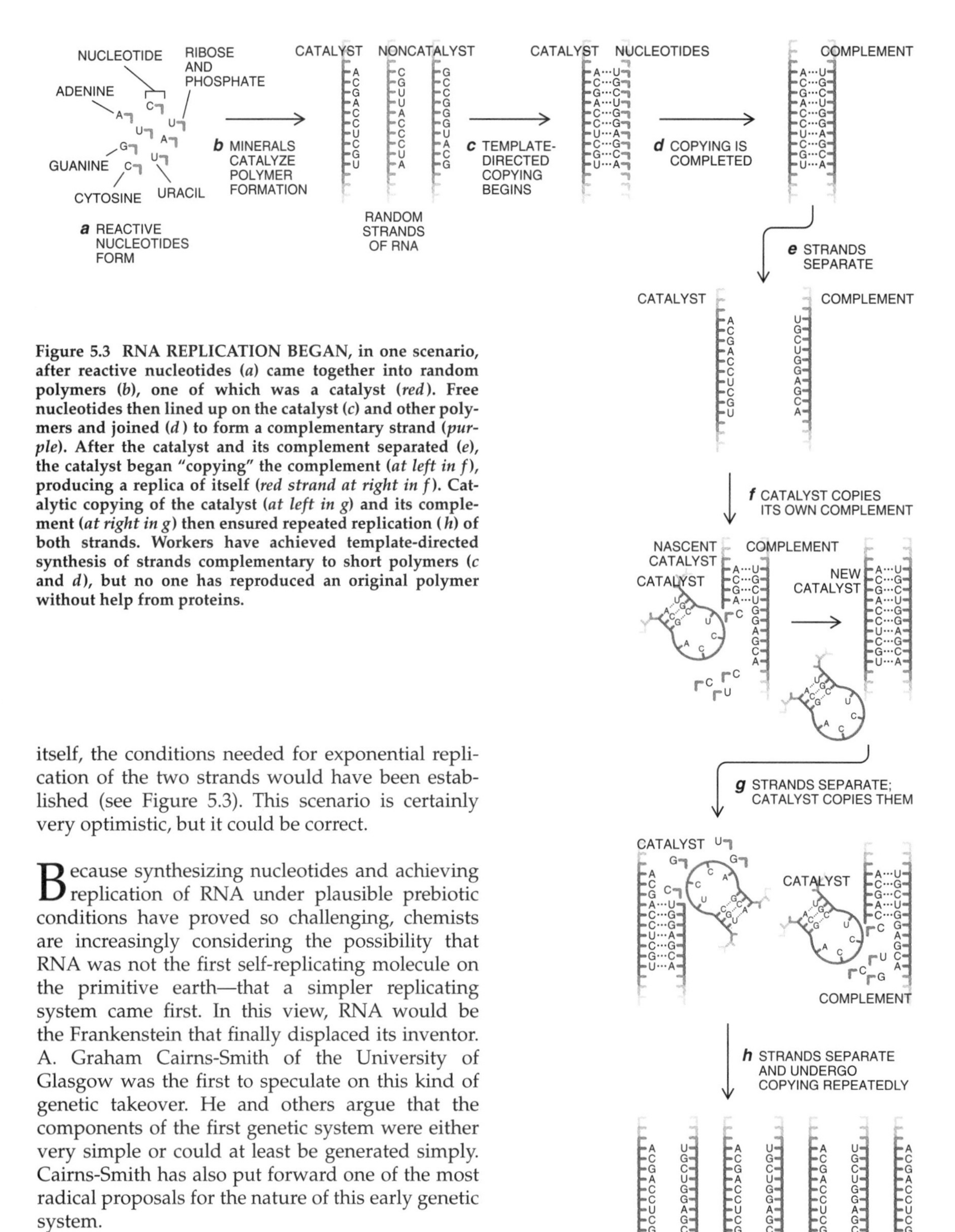

Figure 5.3 RNA REPLICATION BEGAN, in one scenario, after reactive nucleotides (*a*) came together into random polymers (*b*), one of which was a catalyst (*red*). Free nucleotides then lined up on the catalyst (*c*) and other polymers and joined (*d*) to form a complementary strand (*purple*). After the catalyst and its complement separated (*e*), the catalyst began "copying" the complement (*at left in f*), producing a replica of itself (*red strand at right in f*). Catalytic copying of the catalyst (*at left in g*) and its complement (*at right in g*) then ensured repeated replication (*h*) of both strands. Workers have achieved template-directed synthesis of strands complementary to short polymers (*c* and *d*), but no one has reproduced an original polymer without help from proteins.

itself, the conditions needed for exponential replication of the two strands would have been established (see Figure 5.3). This scenario is certainly very optimistic, but it could be correct.

Because synthesizing nucleotides and achieving replication of RNA under plausible prebiotic conditions have proved so challenging, chemists are increasingly considering the possibility that RNA was not the first self-replicating molecule on the primitive earth—that a simpler replicating system came first. In this view, RNA would be the Frankenstein that finally displaced its inventor. A. Graham Cairns-Smith of the University of Glasgow was the first to speculate on this kind of genetic takeover. He and others argue that the components of the first genetic system were either very simple or could at least be generated simply. Cairns-Smith has also put forward one of the most radical proposals for the nature of this early genetic system.

Some 30 years ago he proposed that the very first replicating system was inorganic. He envisaged irregularities in the structure of a clay—for example, an irregular distribution of cations (positively charged ions)—as the repository of genetic information. Replication would be achieved in this example if any given arrangement of the cations in a preformed layer of clay directed the synthesis of a new layer with an almost identical distribution of cations. Selection could be achieved if the distribution of cations in a layer determined how efficiently that layer would be copied. So far no one has tested this daring hypothesis in the laboratory. On theoretical grounds, however, it seems implausible. Structural irregularities in clay that were complicated enough to set the stage for the emergence of RNA probably would not be amenable to accurate self-replication.

Other investigators have also begun to take up the search for alternative genetic materials. In one intriguing example, Eschenmoser has created a molecule called pyranosyl RNA (pRNA) that is closely related to RNA but incorporates a different version of ribose (see Figure 5.4). In natural RNA, ribose contains a five-member ring of four carbon atoms and one oxygen atom; the ribose in Eschenmoser's structure is rearranged to contain an extra carbon atom in the ring.

Eschenmoser finds that complementary strands of pyranosyl RNA can combine by standard Watson-Crick pairing to give double-strand units that permit fewer unwanted variations in structure than are possible with normal RNA. In addition, the strands do not twist around each other, as they do in double-strand RNA. In a world without protein enzymes, twisting could prevent the strands from separating cleanly in preparation for replication. In many ways, then, pyranosyl RNA seems better suited for replication than RNA itself. If simple means for synthesizing ribonucleotides containing a six-member sugar ring were found, a case could be made that this form of RNA may have preceded the more familiar form of the molecule.

In quite a different approach, Peter E. Nielsen of the University of Copenhagen has used computer-assisted model building to design a polymer that combines a proteinlike backbone with nucleic acid bases for side chains. As is true of RNA, one strand of this polymer, or peptide nucleic acid (PNA), can combine stably with a complementary strand; this result implies that, as is true of standard RNA, peptide RNA may be able to serve as a template for the construction of its complement. Many polymers with related backbones may behave in a similar way; perhaps one of them was involved in an early genetic system.

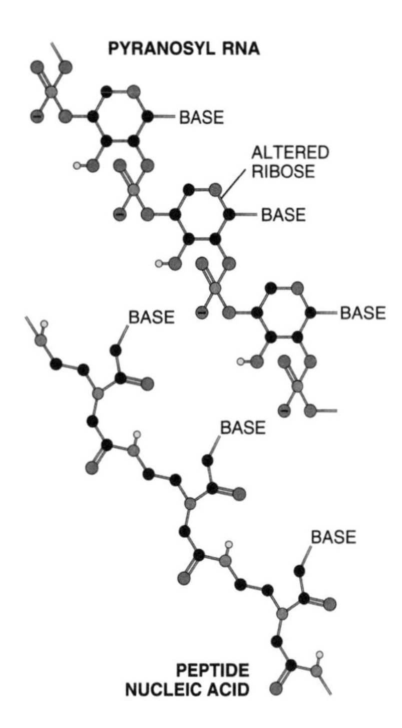

Figure 5.4 TWO MOLECULES related to RNA have been constructed (hydrogen atoms on carbon atoms are not shown). Pyranosyl RNA (*top*) differs from RNA in that its ribose contains a six-member, instead of five-member, ring. Peptide nucleic acid (*bottom*) has nucleic acid bases but a proteinlike backbone. In prebiotic times polymers with similar features might have formed and replicated more readily than did RNA; RNA might have evolved from one such molecule.

Both pyranosyl RNA and peptide nucleic acids rely on Watson-Crick base pairs as the structural element that makes complementary pairing possible. Investigators interested in discovering simpler genetic systems are also trying to build complementary molecules that do not depend on nucleotide bases for template-directed copying. So far, however, there is no good evidence that polymers constructed from such building blocks can replicate. The search for antecedents of RNA can be expected to become a major focus of experimentation for prebiotic chemists.

Whether RNA arose spontaneously or replaced some earlier genetic system, its development was

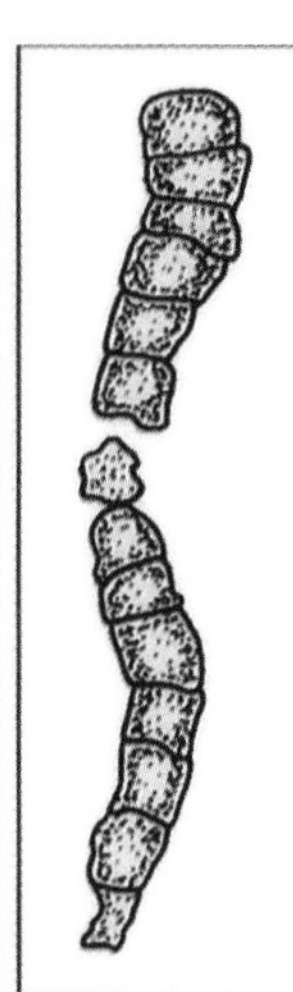

Figure 5.5 MODERN STROMATOLITES (*left photograph*), structures built by cyanobacteria (blue-green algae), grace Shark Bay, Australia. Elsewhere in Australia, J. William Schopf of the University of California at Los Angeles has found remnants of 3.6-billion-year-old stromatolites lying near fossils of 3.5-billion-year-old cells that resemble modern cyanobacteria; the fossils (*right photograph*) apparently represent strings of microscopic cells (*diagram*). Schopf's discoveries indicate that, regardless of how life got started, it was well established within just a billion years after the earth first formed.

probably the watershed event in the development of life (see Figure 5.5). It very likely led to the synthesis of proteins, the formation of DNA and the emergence of a cell that became life's last common ancestor. The precise events giving rise to the RNA world remain unclear. As we have seen, investigators have proposed many hypotheses, but evidence in favor of each of them is fragmentary at best. The full details of how the RNA world, and life, emerged may not be revealed in the near future. Nevertheless, as chemists, biochemists and molecular biologists cooperate on ever more ingenious experiments, they are sure to fill in many missing parts of the puzzle.

The Evolution of Life on the Earth

The history of life is not necessarily progressive; it is certainly not predictable. The earth's creatures have evolved through a series of contingent and fortuitous events.

• • •

Stephen Jay Gould

Some creators announce their inventions with grand éclat. God proclaimed, "Fiat lux," and then flooded his new universe with brightness. Others bring forth great discoveries in a modest guise, as did Charles Darwin in defining his new mechanism of evolutionary causality in 1859: "I have called this principle, by which each slight variation, if useful, is preserved, by the term Natural Selection."

Natural selection is an immensely powerful yet beautifully simple theory that has held up remarkably well, under intense and unrelenting scrutiny and testing, for 135 years. In essence, natural selection locates the mechanism of evolutionary change in a "struggle" among organisms for reproductive success, leading to improved fit of populations to changing environments. (Struggle is often a metaphorical description and need not be viewed as overt combat, guns blazing. Tactics for reproductive success include a variety of nonmartial activities such as earlier and more frequent mating or better cooperation with partners in raising offspring.) Natural selection is therefore a principle of local adaptation, not of general advance or progress (see Figure 6.1).

Yet powerful though the principle may be, natural selection is not the only cause of evolutionary change (and may, in many cases, be overshadowed by other forces). This point needs emphasis because the standard misapplication of evolutionary theory assumes that biological explanation may be equated with devising accounts, often speculative and conjectural in practice, about the adaptive value of any given feature in its original environment (human aggression as good for hunting, music and religion as good for tribal cohesion, for example). Darwin himself strongly emphasized the multifactorial nature of evolutionary change and warned against too exclusive a reliance on natural selection, by placing the following statement in a maximally conspicuous place at the very end of his introduction: "I am convinced that Natural Selection has been the most important, but not the exclusive, means of modification."

Natural selection is not fully sufficient to explain evolutionary change for two major reasons. First, many other causes are powerful, particularly at levels of biological organization both above and below the traditional Darwinian focus on organisms and their struggles for reproductive success. At the lowest level of substitution in individual base pairs of DNA, change is often effectively neutral and therefore random. At higher

levels, involving entire species or faunas, punctuated equilibrium can produce evolutionary trends by selection of species based on their rates of origin and extirpation, whereas mass extinctions wipe out substantial parts of biotas for reasons unrelated to adaptive struggles of constituent species in "normal" times between such events.

Second, and the focus of this article, no matter how adequate our general theory of evolutionary change, we also yearn to document and understand the actual pathway of life's history. Theory, of course, is relevant to explaining the pathway (nothing about the pathway can be inconsistent with good theory, and theory can predict certain general aspects of life's geologic pattern). But the actual pathway is strongly *underdetermined* by our general theory of life's evolution. This point needs some belaboring as a central yet widely misunderstood aspect of the world's complexity. Webs and chains of historical events are so intricate, so imbued with random and chaotic elements, so unrepeatable in encompassing such a multitude of unique (and uniquely interacting) objects, that standard models of simple prediction and replication do not apply.

History can be explained, with satisfying rigor if evidence be adequate, after a sequence of events unfolds, but it cannot be predicted with any precision beforehand. Pierre-Simon Laplace, echoing the growing and confident determinism of the late 18th century, once said that he could specify all future states if he could know the position and motion of all particles in the cosmos at any moment, but the nature of universal complexity shatters this chimerical dream. History includes too much chaos, or extremely sensitive dependence on minute and unmeasurable differences in initial conditions, leading to massively divergent outcomes based on tiny and unknowable disparities in starting points. And history includes too much contingency, or shaping of present results by long chains of unpredictable antecedent states, rather than immediate determination by timeless laws of nature.

Homo sapiens did not appear on the earth, just a geologic second ago, because evolutionary theory predicts such an outcome based on themes of progress and increasing neural complexity. Humans arose, rather, as a fortuitous and contingent outcome of thousands of linked events, any one of which could have occurred differently and sent history on an alternative pathway that would not have led to consciousness. To cite just four among a multitude: (1) If our inconspicuous and fragile lineage had not been among the few survivors of the initial radiation of multicellular animal life in the Cambrian explosion 530 million years ago, then no vertebrates would have inhabited the earth at all. (Only one member of our chordate phylum, the genus *Pikaia*, has been found among these earliest fossils. This small and simple swimming creature, showing its allegiance to us by possessing a notochord, or dorsal stiffening rod, is among the rarest fossils of the Burgess Shale, our best preserved Cambrian fauna.) (2) If a small and unpromising group of lobe-finned fishes had not evolved fin bones with a strong central axis capable of bearing weight on land, then vertebrates might never have become terrestrial. (3) If a large extraterrestrial body had not struck the earth 65 million years ago, then dinosaurs would still be dominant and mammals insignificant (the situation that had prevailed for 100 million years previously). (4) If a small lineage of primates had not evolved upright posture on the drying African savannas just two to four million years ago, then our ancestry might have ended in a line of apes that, like the chimpanzee and gorilla today, would have become ecologically marginal and probably doomed to extinction despite their remarkable behavioral complexity.

Therefore, to understand the events and generalities of life's pathway, we must go beyond principles of evolutionary theory to a paleontological examination of the contingent pattern of life's history on our planet—the single actualized version among millions of plausible alternatives that happened not to occur. Such a view of life's history is highly contrary both to conventional deterministic models of Western science and to the deepest social traditions and psychological hopes of Western culture for a history culminating in humans as life's highest expression and intended planetary steward.

Figure 6.1 SLAB CONTAINING SPECIMENS of *Pteridinium* from Namibia shows a prominent organism from the earth's first multicellular fauna, called Ediacaran, which appeared some 600 million years ago. The Ediacaran animals died out before the Cambrian explosion of modern life. These thin, quilted, sheetlike organisms may be ancestral to some modern forms but may also represent a separate and ultimately failed experiment in multicellular life. The history of life tends to move in quick and quirky episodes, rather than by gradual improvement.

Science can, and does, strive to grasp nature's factuality, but all science is socially embedded, and all scientists record prevailing "certainties," however hard they may be aiming for pure objectivity. Darwin himself, in the closing lines of *The Origin of Species*, expressed Victorian social preference more than nature's record in writing: "As natural selection works solely by and for the good of each being, all corporeal and mental endowments will tend to progress towards perfection."

Life's pathway certainly includes many features predictable from laws of nature, but these aspects are too broad and general to provide the "rightness" that we seek for validating evolution's particular results—roses, mushrooms, people and so forth. Organisms adapt to, and are constrained by, physical principles. It is, for example, scarcely surprising, given laws of gravity, that the largest vertebrates in the sea (whales) exceed the heaviest animals on land (elephants today, dinosaurs in the past), which, in turn, are far bulkier than the largest vertebrate that ever flew (extinct pterosaurs of the Mesozoic era).

Predictable ecological rules govern the structuring of communities by principles of energy flow and thermodynamics (more biomass in prey than in predators, for example). Evolutionary trends, once started, may have local predictability ("arms races," in which both predators and prey hone their defenses and weapons, for example—a pattern that Geerat J. Vermeij of the University of California at Davis has called "escalation" and documented in increasing strength of both crab claws and shells of their gastropod prey through time). But laws of nature do not tell us why we have crabs and snails at all, why insects rule the multicellular world and why vertebrates rather than persistent algal mats exist as the most complex forms of life on the earth.

Relative to the conventional view of life's history as an at least broadly predictable process of gradually advancing complexity through time, three features of the paleontological record stand out in opposition and shall therefore serve as organizing themes for the rest of this chapter: the constancy of modal complexity throughout life's history; the concentration of major events in short bursts interspersed with long periods of relative stability; and the role of external impositions, primarily mass extinctions, in disrupting patterns of "normal" times. These three features, combined with more general themes of chaos and contingency, require a new framework for conceptualizing and drawing life's history, and this chapter therefore closes with suggestions for a different iconography of evolution.

The primary paleontological fact about life's beginnings points to predictability for the onset and very little for the particular pathways thereafter. The earth is 4.6 billion years old, but the oldest rocks date to about 3.9 billion years because the earth's surface became molten early in its history, a result of bombardment by large amounts of cosmic debris during the solar system's coalescence, and of heat generated by radioactive decay of short-lived isotopes. These oldest rocks are too metamorphosed by subsequent heat and pressure to preserve fossils (though some scientists interpret the proportions of carbon isotopes in these rocks as signs of organic production). The oldest rocks sufficiently unaltered to retain cellular fossils—African and Australian sediments dated to 3.5 billion years old—do preserve prokaryotic cells (bacteria and cyanophytes) and stromatolites (mats of sediment trapped and bound by these cells in shallow marine waters). Thus, life on the earth evolved quickly and is as old as it could be. This fact alone seems to indicate an inevitability, or at least a predictability, for life's origin from the original chemical constituents of atmosphere and ocean.

No one can doubt that more complex creatures arose sequentially after this prokaryotic beginning—first eukaryotic cells, perhaps about two billion years ago, then multicellular animals about 600 million years ago, with a relay of highest complexity among animals passing from invertebrates, to marine vertebrates and, finally (if we wish, albeit parochially, to honor neural architecture as a primary criterion), to reptiles, mammals and humans. This is the conventional sequence represented in the old charts and texts as an "age of invertebrates," followed by an "age of fishes," "age of reptiles," "age of mammals," and "age of man" (to add the old gender bias to all the other prejudices implied by this sequence).

I do not deny the facts of the preceding paragraph but wish to argue that our conventional desire to view history as progressive, and to see humans as predictably dominant, has grossly distorted our interpretation of life's pathway by falsely placing in the center of things a relatively minor phenomenon that arises only as a side consequence of a physically constrained starting point. The most salient feature of life has been the stability of its bacterial mode from the beginning of the fossil record until today and, with little doubt, into all future time so long as the earth endures. This is truly the "age of bacteria"—as it was in the beginning, is now and ever shall be.

For reasons related to the chemistry of life's origin and the physics of self-organization, the first living things arose at the lower limit of life's conceivable, preservable complexity. Call this lower limit the "left wall" for an architecture of complexity (see Figure 6.2). Since so little space exists between the left wall and life's initial bacterial mode in the fossil record, only one direction for future increment exists—toward greater complexity at the right. Thus, every once in a while, a more complex creature evolves and extends the range of life's diversity in the only available direction. In technical terms, the distribution of complexity becomes more strongly right skewed through these occasional additions.

But the additions are rare and episodic. They do not even constitute an evolutionary series but form a motley sequence of distantly related taxa, usually depicted as eukaryotic cell, jellyfish, trilobite, nautiloid, eurypterid (a large relative of horseshoe crabs), fish, an amphibian such as *Eryops*, a dinosaur, a mammal and a human being. This sequence cannot be construed as the major thrust or trend of life's history. Think rather of an occasional creature tumbling into the empty right region of complexity's space. Throughout this entire time, the bacterial mode has grown in height and remained constant in position. Bacteria represent the great success story of life's pathway. They occupy a wider domain of environments and span a broader range of biochemistries than any other group. They are adaptable, indestructible and astoundingly diverse. We cannot even imagine

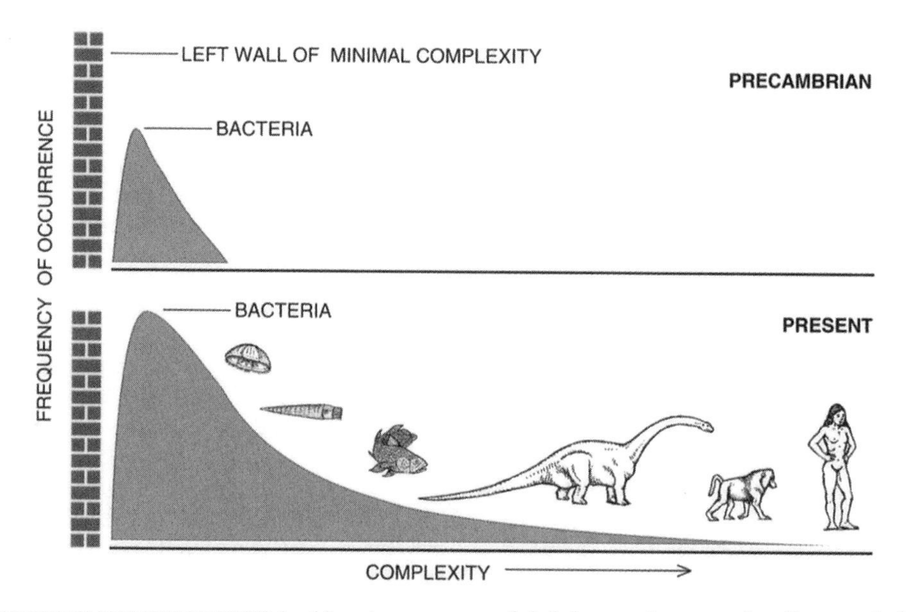

Figure 6.2 PROGRESS DOES NOT RULE (and is not even a primary thrust of) the evolutionary process. For reasons of chemistry and physics, life arises next to the "left wall" of its simplest conceivable and preservable complexity. This style of life (bacterial) has remained most common and most successful. A few creatures occasionally move to the right, thus extending the right tail in the distribution of complexity. Many always move to the left, but they are absorbed within space already occupied. Note that the bacterial mode has never changed in position, but just grown higher.

how anthropogenic intervention might threaten their extinction, although we worry about our impact on nearly every other form of life. The number of *Escherichia coli* cells in the gut of each human being exceeds the number of humans that has ever lived on this planet.

One might grant that complexification for life as a whole represents a pseudotrend based on constraint at the left wall but still hold that evolution within particular groups differentially favors complexity when the founding lineage begins far enough from the left wall to permit movement in both directions. Empirical tests of this interesting hypothesis are just beginning (as concern for the subject mounts among paleontologists), and we do not yet have enough cases to advance a generality. But the first two studies—by Daniel W. McShea of the University of Michigan on mammalian vertebrae and by George F. Boyajian of the University of Pennsylvania on ammonite suture lines—show no evolutionary tendencies to favor increased complexity.

Moreover, when we consider that for each mode of life involving greater complexity, there probably exists an equally advantageous style based on greater simplicity of form (as often found in parasites, for example), then preferential evolution toward complexity seems unlikely a priori. Our impression that life evolves toward greater complexity is probably only a bias inspired by parochial focus on ourselves, and consequent over-attention to complexifying creatures, while we ignore just as many lineages adapting equally well by becoming simpler in form. The morphologically degenerate parasite, safe within its host, has just as much prospect for evolutionary success as its gorgeously elaborate relative coping with the slings and arrows of outrageous fortune in a tough external world.

Even if complexity is only a drift away from a constraining left wall, we might view trends in this direction as more predictable and characteristic of life's pathway as a whole if increments of complexity accrued in a persistent and gradually accumulating manner through time. But nothing about life's history is more peculiar with respect to this common (and false) expectation than the actual pattern of extended stability and rapid episodic movement, as revealed by the fossil record.

Life remained almost exclusively unicellular for the first five sixths of its history—from the first recorded fossils at 3.5 billion years to the first well-documented multicellular animals less than 600 million years ago. (Some simple multicellular algae evolved more than a billion years ago, but these organisms belong to the plant kingdom and have no genealogical connection with animals.) This long period of unicellular life does include, to be sure, the vitally important transition from simple prokaryotic cells without organelles to eukaryotic cells with nuclei, mitochondria and other complexities of intracellular architecture—but no recorded attainment of multicellular animal organization for a full three billion years. If complexity is such a good thing, and multicellularity represents its initial phase in our usual view, then life certainly took its time in making this crucial step. Such delays speak strongly against general progress as the major theme of life's history, even if they can be plausibly explained by lack of sufficient atmospheric oxygen for most of Precambrian time or by failure of unicellular life to achieve some structural threshold acting as a prerequisite to multicellularity.

More curiously, all major stages in organizing animal life's multicellular architecture then occurred in a short period beginning less than 600 million years ago and ending by about 530 million years ago—and the steps within this sequence are also discontinuous and episodic, not gradually accumulative. The first fauna, called Ediacaran to honor the Australian locality of its initial discovery but now known from rocks on all continents, consists of highly flattened fronds, sheets and circlets composed of numerous slender segments quilted together. The nature of the Ediacaran fauna is now a subject of intense discussion. These creatures do not seem to be simple precursors of later forms. They may constitute a separate and failed experiment in animal life, or they may represent a full range of diploblastic (two-layered) organization, of which the modern phylum Cnidaria (corals, jellyfishes and their allies) remains as a small and much altered remnant.

In any case, they apparently died out well before the Cambrian biota evolved. The Cambrian then began with an assemblage of bits and pieces, frustratingly difficult to interpret, called the "small shelly fauna." The subsequent main pulse, starting about 530 million years ago, constitutes the famous Cambrian explosion (see Figure 6.3), during which all but one modern phylum of animal life made a first appearance in the fossil record. (Geologists had previously allowed up to 40 million years for this event, but an elegant study, published in 1993, clearly restricts this period of phyletic flowering to a mere five million years.) The Bryozoa, a group of sessile and colonial marine organisms, do not arise until the beginning of the subsequent, Ordovician period, but this apparent delay may be an artifact of failure to discover Cambrian representatives.

Although interesting and portentous events have occurred since, from the flowering of dinosaurs to the origin of human consciousness, we do not exaggerate greatly in stating that the subsequent history of animal life amounts to little more than variations on anatomical themes established during the Cambrian explosion within five million years. Three billion years of unicellularity, followed by five million years of intense creativity and then capped by more than 500 million years of variation on set anatomical themes can scarcely be read as a predictable, inexorable or continuous trend toward progress or increasing complexity.

We do not know why the Cambrian explosion could establish all major anatomical designs so quickly. An "external" explanation based on ecology seems attractive: the Cambrian explosion represents an initial filling of the "ecological barrel" of niches for multicellular organisms, and any experiment found a space. The barrel has never emptied since; even the great mass extinctions left a few species in each principal role, and their occupation of ecological space forecloses opportunity for fundamental novelties. But an "internal" explanation based on genetics and development also seems necessary as a complement: the earliest multicellular animals may have maintained a flexibility for genetic change and embryological transformation that became greatly reduced as organisms "locked in" to a set of stable and successful designs.

In any case, this initial period of both internal and external flexibility yielded a range of invertebrate anatomies (see Figure 6.4) that may have exceeded (in just a few million years of production) the full scope of animal form in all the earth's environments today (after more than 500 million years of additional time for further expansion). Scientists are divided on this question. Some claim that the anatomical range of this initial explosion exceeded that of modern life, as many early experiments died out and no new phyla have ever arisen. But scientists most strongly opposed to this view allow that Cambrian diversity at least equaled the modern range—so even the most cautious opinion holds that 500 million subsequent years of opportunity have not expanded the Cambrian range, achieved in just five million years. The Cambrian explosion was the most remarkable and puzzling event in the history of life.

Moreover, we do not know why most of the early experiments died, while a few survived to become our modern phyla. It is tempting to say that the victors won by virtue of greater anatomical complexity, better ecological fit or some other predictable feature of conventional Darwinian struggle. But no recognized traits unite the victors, and the radical alternative must be entertained that each early experiment received little more than the equivalent of a ticket in the largest lottery ever played out on our planet—and that each surviving lineage, including our own phylum of vertebrates, inhabits the earth today more by the luck of the draw than by any predictable struggle for existence. The history of multicellular animal life may be more a story of great reduction in initial possibilities, with stabilization of lucky survivors, than a conventional tale of steady ecological expansion and morphological progress in complexity.

Finally, this pattern of long stasis, with change concentrated in rapid episodes that establish new equilibria, may be quite general at several scales of time and magnitude, forming a kind of fractal pattern in self-similarity. According to the punctuated equilibrium model of speciation, trends within lineages occur by accumulated episodes of geologically instantaneous speciation, rather than by gradual change within continuous populations (like climbing a staircase rather than rolling a ball up an inclined plane).

Even if evolutionary theory implied a potential internal direction for life's pathway (although previous facts and arguments in this article cast doubt on such a claim), the occasional imposition of a rapid and substantial, perhaps even truly catastrophic, change in environment would have intervened to stymie the pattern. These environmental changes trigger mass extinction of a high

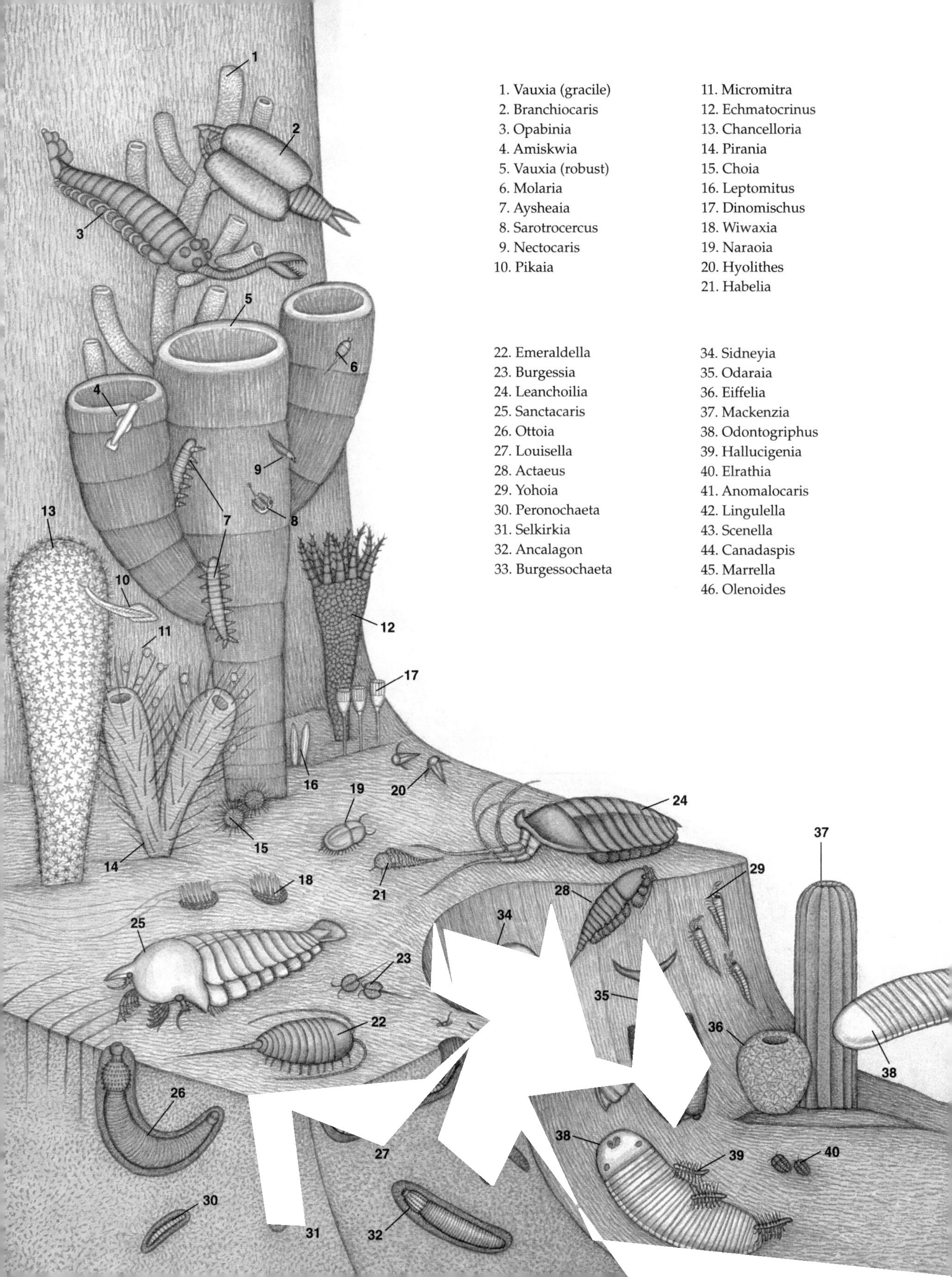

1. Vauxia (gracile)
2. Branchiocaris
3. Opabinia
4. Amiskwia
5. Vauxia (robust)
6. Molaria
7. Aysheaia
8. Sarotrocercus
9. Nectocaris
10. Pikaia
11. Micromitra
12. Echmatocrinus
13. Chancelloria
14. Pirania
15. Choia
16. Leptomitus
17. Dinomischus
18. Wiwaxia
19. Naraoia
20. Hyolithes
21. Habelia
22. Emeraldella
23. Burgessia
24. Leanchoilia
25. Sanctacaris
26. Ottoia
27. Louisella
28. Actaeus
29. Yohoia
30. Peronochaeta
31. Selkirkia
32. Ancalagon
33. Burgessochaeta
34. Sidneyia
35. Odaraia
36. Eiffelia
37. Mackenzia
38. Odontogriphus
39. Hallucigenia
40. Elrathia
41. Anomalocaris
42. Lingulella
43. Scenella
44. Canadaspis
45. Marrella
46. Olenoides

percentage of the earth's species and may so derail any internal direction and so reset the pathway that the net pattern of life's history looks more capricious and concentrated in episodes than steady and directional. Mass extinctions have been recognized since the dawn of paleontology; the major divisions of the geologic time scale were established at boundaries marked by such events. But until the revival of interest that began in the late 1970s, most paleontologists treated mass extinctions only as intensifications of ordinary events, leading (at most) to a speeding up of tendencies that pervaded normal times. In this gradualistic theory of mass extinction, these events really took a few million years to unfold (with the appearance of suddenness interpreted as an artifact of an imperfect fossil record), and they only made the ordinary occur faster (more intense Darwinian competition in tough times, for example, leading to even more efficient replacement of less adapted by superior forms).

The reinterpretation of mass extinctions as central to life's pathway and radically different in effect began with the presentation of data by Luis and Walter Alvarez in 1979, indicating that the impact of a large extraterrestrial object (they sug-

Figure 6.3 GREAT DIVERSITY quickly evolved at the dawn of multicellular animal life during the Cambrian period (530 million years ago). The creatures shown here are all found in the Middle Cambrian Burgess Shale fauna of Canada. They include some familiar forms (sponges, brachiopods) that have survived. But many creatures (such as the giant *Anomalocaris,* at the lower right, largest of all the Cambrian animals) did not live for long and are so anatomically peculiar (relative to survivors) that we cannot classify them among known phyla.

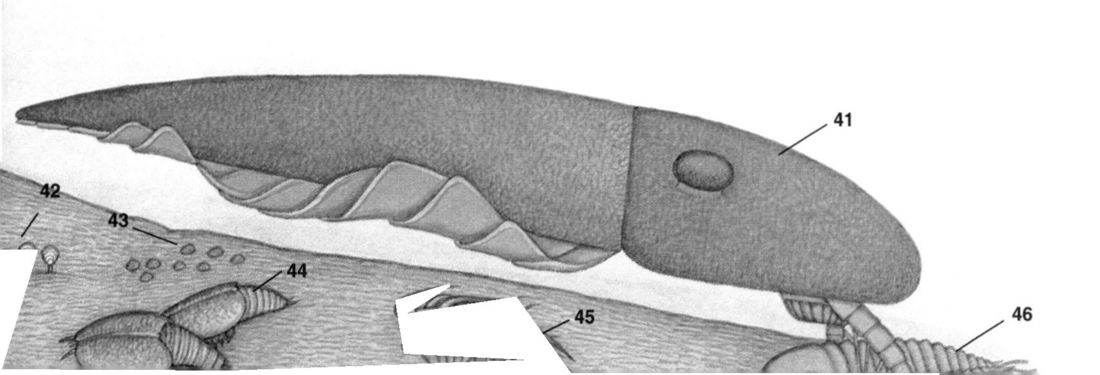

Figure 6.4 NEW ICONOGRAPHY OF LIFE'S TREE shows that maximal diversity in anatomical forms (not in number of species) is reached very early in life's multicellular history. Later times feature extinction of most of these initial experiments and enormous success within surviving lines. This success is measured in the proliferation of species but not in the development of new anatomies. Today we have more species than ever before, although they are restricted to fewer basic anatomies.

gested an asteroid seven to 10 kilometers in diameter) set off the last great extinction at the Cretaceous-Tertiary boundary 65 million years ago. Although the Alvarez hypothesis initially received very skeptical treatment from scientists (a proper approach to highly unconventional explanations), the case now seems virtually proved by discovery of the "smoking gun," a crater of appropriate size and age located off the Yucatán peninsula in Mexico.

This reawakening of interest also inspired paleontologists to tabulate the data of mass extinction more rigorously. Work by David M. Raup, J. J. Sepkoski, Jr., and David Jablonski of the University of Chicago has established that multicellular animal life experienced five major (end of Ordovician, late Devonian, end of Permian, end of Triassic and end of Cretaceous) and many minor mass extinctions during its 530-million-year history. We have no clear evidence that any but the last of these events was triggered by catastrophic impact, but such careful study leads to the general conclusion that mass extinctions were more frequent, more rapid, more extensive in magnitude and more different in effect than paleontologists had previously realized. These four properties encompass the radical implications of mass extinction for understanding life's pathway as more contingent and chancy than predictable and directional.

Mass extinctions are not random in their impact on life. Some lineages succumb and others survive as sensible outcomes based on presence or absence of evolved features. But especially if the triggering cause of extinction be sudden and catastrophic, the reasons for life or death may be random with respect to the original value of key features when first evolved in Darwinian struggles of normal times. This "different rules" model of mass extinction imparts a quirky and unpredictable character to life's pathway based on the evident claim that lineages cannot anticipate future contingencies of such magnitude and different operation.

To cite two examples from the impact-triggered Cretaceous-Tertiary extinction 65 million years ago: First, an important study published in 1986 noted that diatoms survived the extinction far better than other single-celled plankton (primarily coccoliths and radiolaria). This study found that many diatoms had evolved a strategy of dormancy by encystment, perhaps to survive through seasonal periods of unfavorable conditions (months of darkness in polar species as otherwise fatal to these photosynthesizing cells; sporadic availability of silica needed to construct their skeletons). Other planktonic cells had not evolved any mechanisms for dormancy. If the terminal Cretaceous impact produced a dust cloud that blocked light for several months or longer (one popular idea for a "killing scenario" in the extinction), then diatoms may have survived as a fortuitous result of dormancy mechanisms evolved for the entirely different function of weathering seasonal droughts in ordinary times. Diatoms are not superior to radiolaria or other plankton that succumbed in far greater numbers; they were simply fortunate to possess a favorable feature, evolved for other reasons, that fostered passage through the impact and its sequelae.

Second, we all know that dinosaurs perished in the end Cretaceous event and that mammals therefore rule the vertebrate world today. Most people assume that mammals prevailed in these tough times for some reason of general superiority over dinosaurs. But such a conclusion seems most unlikely. Mammals and dinosaurs had coexisted for 100 million years, and mammals had remained rat-sized or smaller, making no evolutionary "move" to oust dinosaurs. No good argument for mammalian prevalence by general superiority has ever been advanced, and fortuity seems far more likely. As one plausible argument, mammals may have survived partly as a result of their small size (with much larger, and therefore extinction-resistant, populations as a consequence, and less ecological specialization with more places to hide, so to speak). Small size may not have been a positive mammalian adaptation at all, but more a sign of inability ever to penetrate the dominant domain of dinosaurs. Yet this "negative" feature of normal times may be the key reason for mammalian survival and a prerequisite to my writing and your reading this chapter today.

Sigmund Freud often remarked that great revolutions in the history of science have but one common, and ironic, feature: they knock human arrogance off one pedestal after another of our previous conviction about our own self-importance. In Freud's three examples, Copernicus moved our home from center to periphery; Darwin then relegated us to "descent from an animal world"; and, finally (in one of the least modest statements of intellectual history), Freud himself discovered the unconscious and exploded the myth of a fully rational mind.

In this wise and crucial sense, the Darwinian revolution remains woefully incomplete because, even though thinking humanity accepts the fact of evolution, most of us are still unwilling to abandon the comforting view that evolution means (or at least embodies a central principle of) progress defined to render the appearance of something like human consciousness either virtually inevitable or at least predictable. The pedestal is not smashed until we abandon progress or complexification as a central principle and come to entertain the strong possibility that *H. sapiens* is but a tiny, late-arising twig on life's enormously arborescent bush—a small bud that would almost surely not appear a second time if we could replant the bush from seed and let it grow again.

Primates are visual animals, and the pictures we draw betray our deepest convictions and display our current conceptual limitations. Artists have always painted the history of fossil life as a sequence from invertebrates, to fishes, to early terrestrial amphibians and reptiles, to dinosaurs, to mammals and, finally, to humans. There are no exceptions; all sequences painted since the inception of this genre in the 1850s follow the convention (see Figure 6.5).

Yet we never stop to recognize the almost absurd biases coded into this universal mode. No scene ever shows another invertebrate after fishes evolved, but invertebrates did not go away or stop evolving! After terrestrial reptiles emerge, no subsequent scene ever shows a fish (later oceanic tableaux depict only such returning reptiles as ichthyosaurs and plesiosaurs). But fishes did not stop evolving after one small lineage managed to invade the land. In fact, the major event in the evolution of fishes, the origin and rise to dominance of the teleosts, or modern bony fishes, occurred during the time of the dinosaurs and is therefore never shown at all in any of these sequences—even though teleosts include more than half of all species of vertebrates. Why should humans appear at the end of all sequences? Our order of primates is ancient among mammals, and many other successful lineages arose later than we did.

We will not smash Freud's pedestal and complete Darwin's revolution until we find, grasp and accept another way of drawing life's history. J.B.S. Haldane proclaimed nature "queerer than we can suppose," but these limits may only be socially imposed conceptual locks rather then inherent restrictions of our neurology. New icons might

Figure 6.5 CLASSICAL REPRESENTATIONS OF LIFE'S HISTORY reveal the biases of viewing evolution as progress and complexification. These paintings by Charles R. Knight from a 1942 issue of *National Geographic,* first show invertebrates of the Burgess Shale (*panel 1*), but once fishes evolve (*panel 2*), no subsequent scene ever shows another invertebrate, although they did not go away or stop evolving. When land vertebrates arise (*panel 3*), we never see another fish, even though return of land vertebrate lineages to the sea may be depicted (*panel 4*). The sequence always ends with mammals (*panel 5*)—even though fishes, invertebrates and reptiles are still thriving—and, of course, humans (*panel 6*).

break the locks. Trees—or rather copiously and luxuriantly branching bushes—rather than ladders and sequences hold the key to this conceptual transition.

We must learn to depict the full range of variation, not just our parochial perception of the tiny right tail of most complex creatures. We must recognize that this tree may have contained a maximal number of branches near the beginning of multicellular life and that subsequent history is for the most part a process of elimination and lucky survivorship of a few, rather than continuous flowering, progress and expansion of a growing multitude. We must understand that little twigs are contingent nubbins, not predictable goals of the massive bush beneath. We must remember the greatest of all Biblical statements about wisdom: "She is a tree of life to them that lay hold upon her; and happy is every one that retaineth her."

The Search for Extraterrestrial Life

The earth remains the only inhabited world known so far, but scientists are finding that the universe abounds with the chemistry of life.

• • •

Carl Sagan

In the past few decades the human species has begun, seriously and systematically, to look for evidence of life elsewhere. While no one has yet found living organisms beyond the earth, there are some reasons to be encouraged. Robotic space probes have identified worlds where life may once have gained a toehold, even if it does not flourish there today. The *Galileo* spacecraft found clear signs of life during its recent flight past the earth—a reassurance that we really do know how to sniff out at least certain kinds of life. And rapidly accumulating evidence strongly suggests that the universe abounds with planetary systems something like our own.

In practice, the community of scientists concerned with finding life elsewhere in the solar system has contented itself with a chemical approach. Human beings, as well as every other organism on the earth, are based on liquid water and organic molecules. (Organic molecules are carbon-containing compounds other than carbon dioxide and carbon monoxide.) A modest search strategy—looking for necessary if not sufficient criteria—might then begin by looking for liquid water and organic molecules. Of course, such a protocol might miss forms of life about which we are wholly ignorant, but that does not mean we could not detect them by other methods. If a silicon-based giraffe had walked by the *Viking* Mars landers, its portrait would have been taken.

Actually, focusing on organic matter and liquid water is not nearly so parochial and chauvinistic as it might seem. No other chemical element comes close to carbon in the variety and intricacy of the compounds it can form; liquid water provides a superb, stable medium in which organic molecules can dissolve and interact. What is more, organic molecules are surprisingly common in the universe. Astronomers find evidence for them everywhere, from interstellar gas and dust grains to meteorites to many worlds in the outer solar system.

Some other molecules—hydrogen fluoride, for example—might approach water in their ability to dissolve other molecules, but the cosmic abundance of fluorine is extremely low. Certain atoms, such as silicon, might be able to take on some of the roles of carbon in an alternative life chemistry, but the variety of information-bearing molecules they provide seems comparatively sparse. Furthermore, the silicon equivalent of carbon dioxide (silicon dioxide, the major component of ordinary glass) is, on all planetary surfaces, a solid, not a gas. That distinction would certainly complicate the development of a silicon-based metabolism.

On extremely cold worlds, where water is frozen solid, some other solvent—liquid ammonia, for instance—might be a key to a different form of biochemistry. At low temperatures, certain classes of molecules require very little activation energy to undergo chemical reactions, but because our laboratories are at room temperature and not, say, at the temperature of Neptune's satellite Triton, our knowledge of those molecules may well be inadequate. For the moment, though, carbon-and water-based life-forms are the only kinds we know or can even imagine.

On the earth the signature molecules of life are the nucleic acids (DNA and RNA), which constitute the hereditary instructions, and the proteins, which, as enzymes, catalytically control the chemistry of cell and organism. The codebook for translating nucleic acid information into protein structure is essentially identical for all life on the earth. This profound uniformity in the hereditary chemistry suggests that every organism on our planet has evolved from a common instance of the origin of life. If so, we have no way of knowing which aspects of terrestrial life are necessary (required of all living things anywhere) and which are merely contingent (the results of a particular sequence of happenstances that, had they gone otherwise, might have led to organisms having very different properties). We may speculate, but only by examining life elsewhere can biologists truly determine what else is possible (see boxed figure "What Is Life?").

The obvious place to start the search for life is in our own solar system. Robot spacecraft have explored more than 70 planets, satellites, comets and asteroids at distances varying from about 100 to about 100,000 kilometers. These ships have been equipped with magnetometers, charged-particle detectors, imaging systems, and photometric and spectrometric instruments that sense radiation ranging from ultraviolet to kilometer-wavelength

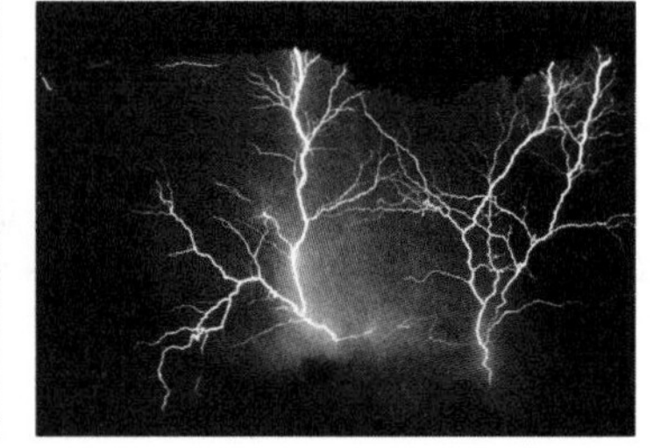

What Is Life?

The search for extraterrestrial life must begin with the question of what we mean by life. "I'll know it when I see it" is an insufficient answer. Some functional definitions are inadequate: one might identify life as anything that ingests, metabolizes and excretes, but this description applies to my car or to a candle flame. Some more sophisticated definitions—for example, life as recognizable by its departure from thermodynamic equilibrium—fall afoul of the circumstance that much of nature (such as lightning and the ozone layer) is out of equilibrium.

Biochemical definitions—for example, defining life in terms of nucleic acids, proteins and other molecules—are clearly chauvinistic. Would we declare an organism that can do everything a bacterium can dead if it was made of very different molecules? The definition that I like best—life is any system capable of reproduction, mutation and reproduction of its mutations—is impractical to apply when we set down a spacecraft on another world: reproduction may not be done in public, and mutations might be comparatively infrequent.

radio. For the moon, Venus and Mars, observations from orbiters and landers have confirmed and expanded on findings transmitted back from flyby spacecraft.

None of these encounters has yielded compelling, or even strongly suggestive, indications of extraterrestrial life. Still, such life, if it exists, might be quite unlike the forms with which we are familiar, or it might be present only marginally. Or the remote-sensing techniques used for examining other worlds might be insensitive to the conceivably subtle signs of life on another world. The most elementary test of these techniques—the detection of life on the earth by an instrumented flyby spacecraft—had, until recently, never been attempted. The National Aeronautics and Space Administration's *Galileo* has rectified that omission.

Galileo is a dual-purpose spacecraft that incorporates a Jupiter orbiter and entry probe; it is currently in interplanetary space and is scheduled to reach the Jupiter system in December 1995. For technical reasons, NASA was unable to send *Galileo* on a direct course to Jupiter; instead the mission incorporated three gravitational assists—two from the earth and one from Venus—to send it on its journey. This looping course greatly lengthened the transit time, but it also permitted the spacecraft to make close-up observations of our planet. Galileo's instruments were not designed for an earth-encounter mission, so circumstance fortuitously arranged a control experiment: a search for life on the earth using a typical modern planetary probe. The results of *Galileo*'s December 1990 encounter with the earth proved quite enlightening.

An observer looking at the data from *Galileo* would immediately notice some unusual facts about the earth. When my co-workers and I examined spectra taken by *Galileo* at near-infrared wavelengths (just slightly longer than red light), we noted a strong dip in brightness at 0.76 micron, a wavelength at which molecular oxygen absorbs radiation. The prominence of the absorption feature implies an enormous abundance of molecular oxygen in the earth's atmosphere, many orders of magnitude greater than is found on any other planet in the solar system.

Oxygen slowly combines with the rocks on the earth's surface, so the oxygen-rich atmosphere requires a replenishing mechanism. Some oxygen is freed when ultraviolet light from the sun splits apart molecules of water (H_2O), and the low-mass hydrogen atoms preferentially escape into space. But the great concentration of oxygen (20 percent) in the earth's dense atmosphere is very hard to explain by this process.

If visible light, rather than ultraviolet, could split water molecules, the abundance of oxygen could be understood, because the sun emits many more photons of visible light than of ultraviolet. But photons of visible light are too feeble to sever the H-OH bond in water. If there were a way to combine two visible light photons to break apart the water molecule, then everything would have a ready solution. Yet so far as we know, there is no way to accomplish this feat—except through life, specifically through photosynthesis in plants. The prevalence of molecular oxygen in the earth's atmosphere is our first clue that the planet bears life.

When *Galileo* photographed the earth (see Figure 7.1), it found unmistakable evidence of a sharp absorption band painting the continents: some substance was soaking up radiation at wavelengths around 0.7 micron (the far red end of the visible spectrum). No known minerals show such a feature, and it is found nowhere else in the solar system. The mystery substance is in fact just the kind of light-absorbing pigment we would expect if visible photons were being added together to break down water and generate molecular oxygen. *Galileo* detected this pigment—which we know as chlorophyll—covering most of the land area of the earth. (Plants appear green precisely because chlorophyll reflects green light but traps the red and blue.) The prevalence of the chlorophyll red band offers a second reason to think that the earth is an inhabited planet.

Galileo's infrared spectrometer also detected a trace amount, about one part per million, of methane. Although that might seem insignificant, it is in startling disequilibrium with all that oxygen. In the earth's atmosphere, methane rapidly oxidizes into water and carbon dioxide. At thermodynamic equilibrium, calculations indicate that not a single molecule of methane should remain. Some unusual processes (which we know to include bacterial metabolism in bogs, rumina and termites) must steadily refresh the methane supply. The profound methane disequilibrium is a third sign of life on the earth.

Finally, *Galileo*'s plasma-wave instrument picked up narrow-band, pulsed, amplitude-modulated

Figure 7.1 LIFE ON THE EARTH betrays its presence in images and measurements made by the *Galileo* spacecraft. This false-color infrared image reveals a mysterious red-absorbing pigment (chlorophyll, which appears orange-brown here) painting the continents. No such pigment is seen anywhere else in the solar system. Spectra indicate that the earth's atmosphere is unusually rich in molecular oxygen and methane. *Galileo* has boosted scientists' confidence that they may be able to spot the telltale signs of life even if it is different from life on the earth.

radio emissions coming from the earth. These signals begin at the frequency at which radio transmissions on the earth's surface are first able to leak through the ionosphere; they look nothing like natural sources of radio waves, such as lightning and the earth's magnetosphere. Such unusual, orderly radio signals strongly suggest the presence of a technological civilization. This is a fourth sign of life and the only one that would not have been apparent to a similar spacecraft flying by the earth anytime within the past two billion years (see boxed figure "Does Intelligent Life Exist on Other Worlds?").

The *Galileo* mission served as a significant control experiment of the ability of remote-sensing spacecraft to detect life at various stages of evolutionary development on other worlds in the solar system. These positive results encourage us that we would be able to spot the telltale signature of life on other worlds. Given that we have found no such evidence, we tentatively conclude that widespread biological activity now exists, among all the bodies of the solar system, only on the earth.

Mars is the nearest planet whose surface we can see. It has an atmosphere, polar ice caps, seasonal changes and a 24-hour day. To generations of scientists, writers and the public at large, Mars seemed the world most likely to sustain extraterrestrial life. But flybys past and orbiters around Mars have found no excess of molecular oxygen, no substances—whatever their nature—enigmatically and profoundly departing from thermodynamic equilibrium, no unexpected surface pigments and no modulated radio emissions. In 1976 NASA set down two *Viking* landers on Mars. I was an experimenter on that mission. The landers were equipped with instruments sensitive enough to detect life even in unpromising deserts and wastelands on the earth (see Figure 7.2).

One experiment measured the gases exchanged between Martian surface samples and the local atmosphere in the presence of organic nutrients carried from the earth. A second experiment brought a wide variety of organic foodstuffs marked by a radioactive tracer, to see if there were life-forms in the Martian soil that ate the food and oxidized it, giving off radioactive carbon dioxide. A third experiment exposed the Martian soil to radioactive carbon dioxide and carbon monoxide to determine if any of it was taken up by microbes.

To the initial astonishment of, I think, all the scientists involved, each of the three *Viking* experiments gave what at first seemed to be positive results. Gases were exchanged; organic matter was oxidized; carbon dioxide was incorporated into the soil.

But there are reasons that these provocative results are not generally thought to provide a convincing argument for life on Mars. The putative

Does Intelligent Life Exist on Other Worlds?

The search for extraterrestrial intelligence is an attempt to use large radio telescopes, sophisticated receivers and modern data analysis to detect hypothetical signals sent our way by advanced civilizations on planets around other stars. Necessarily, there are great uncertainties in selecting the appropriate wavelength, band pass, polarization, time constant and decoding algorithm with which to search for those signals. Nevertheless, radio technology is inexpensive, likely to be discovered early in the evolution of a technological civilization, readily detectable (not just over interplanetary distances, as *Galileo* has done, but over vast interstellar distances) and capable of transmitting enormous amounts of information. The first large-scale, systematic search program, covering a significant fraction of the wavelengths thought optimal for interstellar communication, was initiated by the National Aeronautics and Space Administration on October 12, 1992. Congress canceled the program a year later, but it will soon be resuscitated using private money. Meanwhile some smaller efforts have made provocative findings.

One promising project is the Megachannel Extraterrestrial Array (META), which is led by Paul Horowitz, a physics professor at Harvard University, and funded mainly by the Planetary Society, the largest space interest group in the world. The antenna used for META appears below. After five years of continuous sky survey and two years of follow-up, Horowitz and I found a handful of candidate radio signals that have extremely narrow bandwidths, that do not seem to share the earth's rotation and that cannot be attributed to specific sources of noise or interference. The only trouble is that none of these sources repeats, and in science nonrepeating data are usually not worth much. The tantalizing aspect of the META findings is that the five strongest signals all lie in the plane of the Milky Way. The likelihood that this alignment happens by chance is something like 0.5 percent. We think more comprehensive searches are worth doing.

metabolic processes of Martian microbes occurred under a wide range of conditions: wet and dry, light and dark, cold (only a little above freezing) and hot (almost the normal boiling point of water). Many microbiologists deem it unlikely that Martian microbes would be so capable under such varied conditions. Another strong reason for skepticism is that an additional experiment to look for organic molecules in the Martian soil gave uniformly negative results, even though the instruments could detect such molecules at a sensitivity of around one part per billion. We expected that any life on Mars—as with life on the earth—would be an expression of the chemistry of carbon-based molecules. To find no such molecules at all was daunting for the optimists among the exobiologists.

Figure 7.2 ***VIKING* 2 LANDER scooped up bits of Martian soil and tested them for the presence of life and organic molecules. Despite tantalizing initial results, the *Viking* experiments suggest that Mars is, at present, a dead world. Future missions may search for fossils of organisms that might have lived billions of years ago, when Mars was warmer and wetter.**

The apparent positive results of the life-detection experiments on the two *Viking* landers is now generally attributed to chemicals that oxidize the soil. These chemicals form when solar ultraviolet light irradiates the Martian atmosphere. A handful of *Viking* scientists still wonder whether extremely tough and resilient organisms might exist, so thinly spread over the Martian soil that their organic chemistry could not be detected but their metabolic processes could. Those scientists do not deny the presence of ultraviolet-generated oxidants, but they emphasize that nobody has yet been able to explain fully the *Viking* life-detection results on the basis of oxidants alone. A few researchers have made tentative claims of finding organic matter in a class of meteorites (the SNC meteorites) that are thought to be bits of the Martian surface blasted into space during ancient impacts. More likely, the organic material consists of contaminants that entered the meteorite after its arrival on our world. So far there are no claims of discovering Martian microbes in these rocks from the sky.

For the moment, it is safe to say that *Viking* found no compelling case for life on Mars. No unambiguous signatures of life emerged from four very different, extremely sensitive experiments conducted at two sites 5,000 kilometers apart on a planet where fast winds transport fine particles around the globe. The *Viking* findings suggest that Mars is, today at least, a lifeless planet.

Could Mars have supported life in the distant past? The answer depends very much on how quickly life can arise, a topic about which we remain sadly ignorant. Astronomers are quite certain that, initially, the earth was inhospitable to life because of the collisions of planetesimals, the planetary building blocks that accreted together to form the earth. Early on, the earth was covered by a deep layer of molten rock. After that magma froze, the occasional arrival of large planetesimals would have boiled the oceans and sterilized the earth, if life had already arisen.

Things did not calm down until about 4.0 billion years ago. And yet fossils reveal that by 3.6 billion years ago the earth abounded with microbial life (including large, basketball-size stromatolites, colonies of microorganisms). These early forms of life seem to have been biochemically very adept. Many were photosynthetic, slowly contributing to the earth's bizarre oxygen-rich atmosphere. Manfred Schidlowski of the Max Planck Institute for Chemistry in Mainz has studied carbon isotope ratios preserved in ancient rocks; that work provided (disputed) evidence that life was already flourishing 3.8 billion years ago.

The inferred time available for the origin of life on the earth is thus being squeezed from two directions. According to current knowledge, that amount of time may be as brief as 100 million years. When I first drew attention to this "squeeze"—in 1973, after lunar samples returned by *Apollo* clarified the chronology of impacts on the moon—I argued that the rapidity with which life arose on the earth may imply that it is a likely process. It is dangerous to extrapolate from a single example, but it would be a truly remarkable circumstance if life arose quickly here while on many other, similar worlds, given comparable time, it did not.

Between 4.0 and 3.8 billion years ago, conditions on Mars, too, may have favored the emergence of life. The surface of Mars is covered with evidence of ancient rivers, lakes and perhaps even oceans more than 100 meters deep. The Mars of 4.0 billion years ago was much warmer and wetter than it is today. Taken together, these pieces of information suggest, although they hardly prove, that life may have arisen on ancient Mars as it did on the ancient earth. If so, as Mars evolved from congenial to desolate, life would have held on in the last remaining refugia—perhaps saline lakes or places where the interior heat had melted the permafrost. Most planetary scientists agree that searching for chemical or morphological fossils of ancient life should have high priority in future Martian exploration. Although it is a long shot, searching for life in contemporary Martian oases might also be a productive endeavor.

It is now clear that organic chemistry has run rampant through the solar system and beyond. Mars has two small satellites, Phobos and Deimos, which, because of their dark color, seem to be made of (or at least covered by) organic matter. They are widely thought to be captured asteroids from farther out in the solar system. Indeed, there seems to be a vast population of small worlds covered with organic matter: the so-called C- and D-type asteroids in the main asteroid belt between Jupiter and Mars; the nuclei of comets such as Halley's Comet; and the newly discovered class of asteroids near the outermost planets. In 1986 the European Space Agency's *Giotto* spacecraft flew directly into the cloud of dust surrounding Halley's Comet, revealing that its nucleus may be made of as much as 25 percent organic matter.

A fairly abundant type of meteorite on the earth, known as carbonaceous chondrite, is thought to consist of fragments from C-type asteroids in the main belt. Carbonaceous meteorites contain an organic residue rich in aromatic and other hydrocarbons. Scientists have also identified a number of amino acids (the building blocks of the proteins) and nucleotide bases (the "rungs" of the DNA double helix, which spell out the genetic code).

Asteroidal and cometary fragments plunging into the atmosphere of the early earth carried with them vast stores of organic molecules. Some of these survived the intense heating on entry and therefore may have made a significant material contribution to the origin of life. Impacts would have delivered similar supplies of organic matter, along with water, to other worlds. Those worlds need not be as richly endowed with liquid water as is the earth for critical steps in prebiological chemistry to occur. The water could be found in ponds, in subsurface reservoirs, as thin films on mineral grains or as ice melts formed by impacts.

One of the most fascinating and instructive worlds illustrating prebiological organic chemistry is Saturn's giant moon, Titan (which is as large as the planet Mercury). Here we can see the synthesis of complex organic molecules happening before our eyes. Titan has an atmosphere 10 times as massive as the earth's, composed mainly of molecular nitrogen, along with a few percent to 10 percent methane. When *Voyager 2* approached Titan in 1981, it could not see the surface, because this world is entirely socked in by an opaque, reddish orange haze (see Figure 7.3). The surface temperature is very low, about 94 kelvins, or -179 degrees Celsius. If we can judge from its density (much lower than that of solid rock) and from the composition of nearby worlds, Titan should have a great deal of water ice on and near its surface. A few simple organic molecules—hydrocarbons and nitriles—are found to be minor constituents of Titan's atmosphere.

Ultraviolet light from the sun, charged particles trapped in Saturn's magnetosphere and cosmic rays all bombard Titan's atmosphere and initiate chemical reactions there. When W. Reid Thompson of Cornell University and I considered the effects of ultraviolet irradiation and simulated those of auroral electron bombardment, we found the results agree well with the observed abundances of gaseous organic constituents.

My colleague Bishun N. Khare and I at Cornell simulated the pressure and composition of the appropriate levels in Titan's atmosphere and irradi-

Figure 7.3 SATURN'S MOON TITAN (at the upper right in this *Voyager 2* composite image) is an intriguing natural laboratory for prebiotic chemistry. Titan has a thick atmosphere in which complex organic solids form and rain down onto the surface.

ated the gases with charged particles. The experiment produced a dark, organic solid that we call Titan tholin, from the Greek word for "muddy." When we measure the optical constants of Titan tholin, we find that it beautifully matches the optical constants derived from observations of the Titan haze (see Figure 7.4). No other proposed material comes close.

Organic molecules continually form in the upper atmosphere of Titan and slowly fall out as new tholins are generated in the upper air. If this process has continued over the past four billion years, Titan's surface must be covered by tens, maybe even hundreds, of meters of tholin and other organic products. Moreover, Thompson and I have calculated that over the history of the solar system, a typical location on Titan has something like a 50–50 chance of having experienced centuries of liquid water from the heat released by impacts. When we mix Titan tholin with water in the laboratory, we make amino acids. There are also traces of nucleotide bases, polycyclic aromatic hydrocarbons and a wonderful brew of other compounds. If 100 million years is enough for the origin of life on the

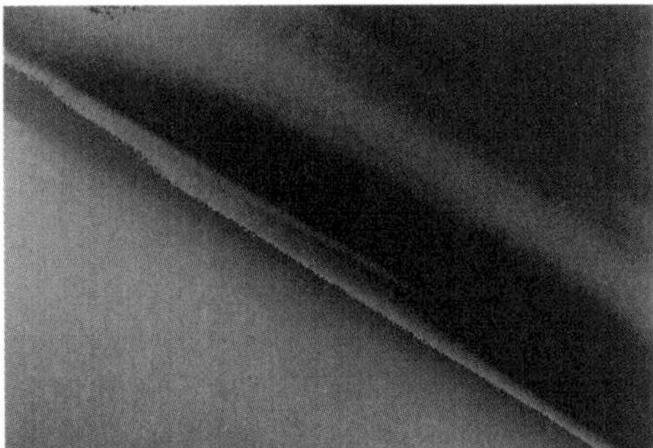

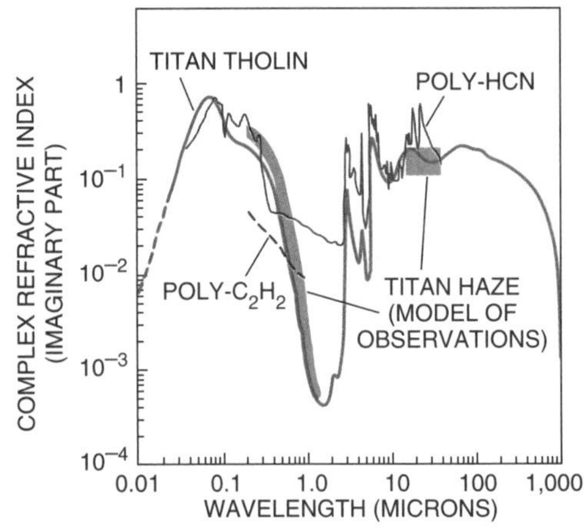

Figure 7.4 LABORATORY SIMULATION of Titan's nitrogen–methane atmosphere (*left*) yields a tarlike accumulation of complex organic molecules, which the author calls Titan tholin. Analogous chemical reactions may give rise to the haze that obscures Titan's surface (*top right*). The optical characteristics of Titan tholin closely match those of Titan's haze (*bottom right*). When combined with liquid water, Titan tholin produces amino acids, nucleotide bases and other molecules important to terrestrial life. Such molecules might have formed in temporary lakes created by cometary impacts on Titan.

earth, could 1,000 years be enough for it on Titan? Could life have started on Titan during the centuries following an impact, when lakes of water or water-ice slurries briefly formed? The first close-up examination of Titan—by a Saturn orbiter and Titan entry probe—is scheduled to occur when the ESA-NASA Cassini mission reaches the Saturn system in about 2004.

When we look beyond our solar system, into the gas and grains that populate interstellar space, again we find striking signs of the prevalence of organic chemistry. Astronomers examining microwaves emitted and absorbed by molecules at distinctive frequencies have identified more than four dozen simple organic compounds in interstellar space—hydrocarbons, amines, alcohols and nitriles, some of them having long, straight carbon chains, such as $HC_{11}N$. When a cloud of interstellar dust grains lies between the earth and some more distant infrared source, it is possible to determine which infrared wavelengths are absorbed by the grains and hence to learn about their composition.

Some of the missing infrared light is widely presumed to have been absorbed by polycyclic aromatics, complex hydrocarbons similar to the compounds found in coal tar. In the part of the infrared spectrum near 3.4 microns, three distinct absorption features are seen. The same patterns appear in the spectra of comets, in tholins made from the irradiation of hydrocarbon ices and in meteoritic organic matter. That infrared fingerprint is probably caused by linked (aliphatic) groups of carbon and hydrogen: $-CH_3$ and $-CH_2$. Yvonne Pendleton and her colleagues at the NASA Ames Research Center find that the best spectral fit seems to be with meteoritic organic matter.

The infrared match among comets, asteroids and interstellar clouds may represent the first direct evidence that asteroids and comets contain organic matter that originated on interstellar grains before gathering together in the infant solar system. But the data are also amenable to an opposite interpretation—that some of the organic matter that formed in the early solar nebula accumulated into asteroids and comets, while some was ejected by the sun into interstellar space. If 100 billion other stars did likewise, they could account for a significant fraction of the organic matter in all the interstellar grains in the galaxy. The prevalence of organic material in the outer solar system, in comets that come from far beyond the outermost planets and in interstellar gas and grains strongly suggests that complex organic matter—relevant to the origin of life—is widely spread throughout the Milky Way.

Organic molecules on bone-dry interstellar grains fried by ultraviolet light and cosmic rays seem an unlikely habitat for the origin of life, however. Life seems to need liquid water, which in turn seems to require planets. Astronomical observations increasingly indicate that planetary systems are common. A surprisingly large number of nearby young stars of roughly solar mass are surrounded by just the kind of disks of gas and dust that scientists going back to Immanuel Kant and Pierre-Simon Laplace, say is needed to explain the origin of the planets in our system. These disks provide a persuasive though still indirect indication that there is a multitude of planets, presumably including earthlike worlds, around other stars.

George W. Wetherill of the Carnegie Institution of Washington has developed detailed models for predicting the distribution of the planets that should be formed in such circumstellar disks. Meanwhile James F. Kasting of Pennsylvania State University has calculated the range of distances from their suns at which planets can support liquid water on their surfaces. Taken together, these two lines of inquiry suggest that a typical planetary system should contain one and maybe even two earthlike planets circling at a distance where liquid water is possible.

Recently Alexander Wolszczan, also at Pennsylvania State, unambiguously detected earthlike planets in a place where most astronomers least expected to find them: around a pulsar, the swiftly spinning neutron-star remnant from a supernova explosion. Based on variations in the timing of radio emissions from the pulsar PSR B1257+12, Wolszczan has deduced the presence of three planets (so far called only A, B and C) orbiting the pulsar (see Figure 7.5).

These worlds are closer to their star than the earth is to ours, and PSR B1257+12 emits in charged particles several times as much energy as does the sun in electromagnetic radiation. If all the charged particles intercepted by A, B and C are transformed into heat, these worlds must almost certainly be too hot for life. But Wolszczan finds hints of at least one additional planet situated farther from the pulsar. For all we know, this superficially unpromising system, 1,400 light-years from the earth, may contain a dark but habitable planet. It is not clear whether these planets survived from before the supernova explosion or, more likely, formed afterward from surrounding debris. Either way, their presence suggests that planetary formation is an unexpectedly common and widespread process.

Numerous searches for planets in infant and mature sunlike systems are under way. The pace of exploration is becoming so quick, and so many new techniques are about to be employed, that it seems likely that in the next few decades astronomers will begin accumulating a sizable inventory of planets around nearby stars.

We have every reason to believe that there are many water-rich worlds something like our own, each provided with a generous complement of complex organic molecules. Those planets that circle sunlike stars could offer environments in which life would have billions of years to arise and evolve. Should not there be an immense number and diversity of inhabited worlds in the Milky Way? Scientists differ about the strength of this argument, but even at its best it is very different from actually detecting life elsewhere. That monumental discovery remains to be made.

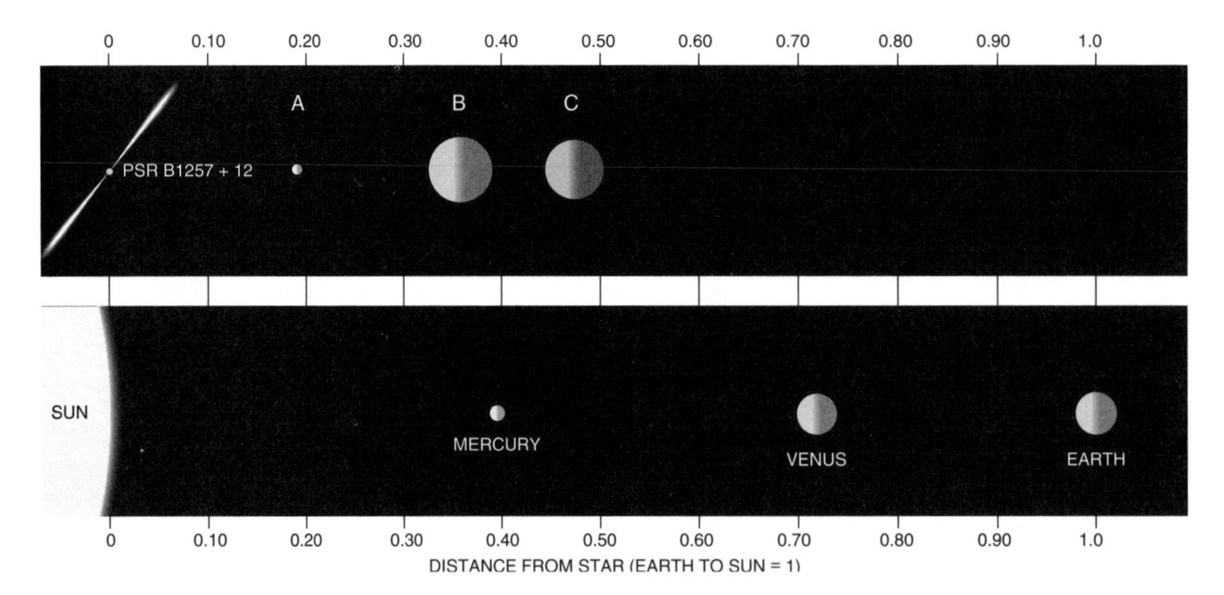

Figure 7.5 EXTRASOLAR PLANETS seem to orbit the star PSR B1257+12, the tiny, dense remnant of an ancient supernova explosion. The spacing of the three surrounding planets—known as A, B and C—resembles the circumstances of our solar system (the sizes of the planets are not drawn to scale). It is possible that a more distant, habitable planet circles this stellar corpse. There is also increasing evidence that planetary systems orbit many sunlike stars, which offer better prospects for life.

The Emergence of Intelligence

Language, foresight, musical skills and other hallmarks of intelligence are connected through an underlying facility that enhances rapid movements.

• • •

William H. Calvin

To most observers, the essence of intelligence is cleverness, a versatility in solving novel problems. Bertrand Russell once wryly noted: "Animals studied by Americans rush about frantically, with an incredible display of hustle and pep, and at last achieve the desired result by chance. Animals observed by Germans sit still and think, and at last evolve the solution out of their inner consciousness." Besides commenting on the scientific fashions of 1927, Russell's remark illustrates the false dichotomy usually made between random trial and error (which intuitively seems unrelated to intelligent behavior) and insight.

Foresight is also said to be an essential aspect of intelligence—particularly after an encounter with one of those terminally clever people who are all tactics and no strategy. Psychologist Jean Piaget emphasized that intelligence was the sophisticated groping that we use when not knowing what to do. Personally, I like the way neurobiologist Horace Barlow of the University of Cambridge frames the issue. He says intelligence is all about making a guess that discovers some new underlying order. This idea neatly covers a lot of ground: finding the solution to a problem or the logic of an argument, happening on an appropriate analogy, creating a pleasing harmony or a witty reply, or guessing what is likely to happen next. Indeed, we all routinely predict what comes next, even when passively listening to a narrative or a melody. That is why a joke's punch line or a P. D. Q. Bach musical parody brings you up short—you were subconsciously predicting something else and are surprised by the mismatch.

We will never agree on a universal definition of intelligence, because it is an open-ended word, like consciousness. Intelligence and consciousness concern the high end of our mental life, but they are frequently confused with more elementary mental processes, such as ones we would use to recognize a friend or tie a shoelace. Of course, such simple neural mechanisms are probably the foundations from which our abilities to handle logic and metaphor evolved.

But how did that occur? That's an evolutionary question and a neurophysiological one as well. Both kinds of answers are needed if we are to understand our own intelligence. They might even help us appreciate how an artificial or an exotic intelligence could evolve.

Did our intelligence arise from having more of what other animals have? The two-millimeter-thick cerebral cortex is the part of the brain most involved with making novel associations. Ours is

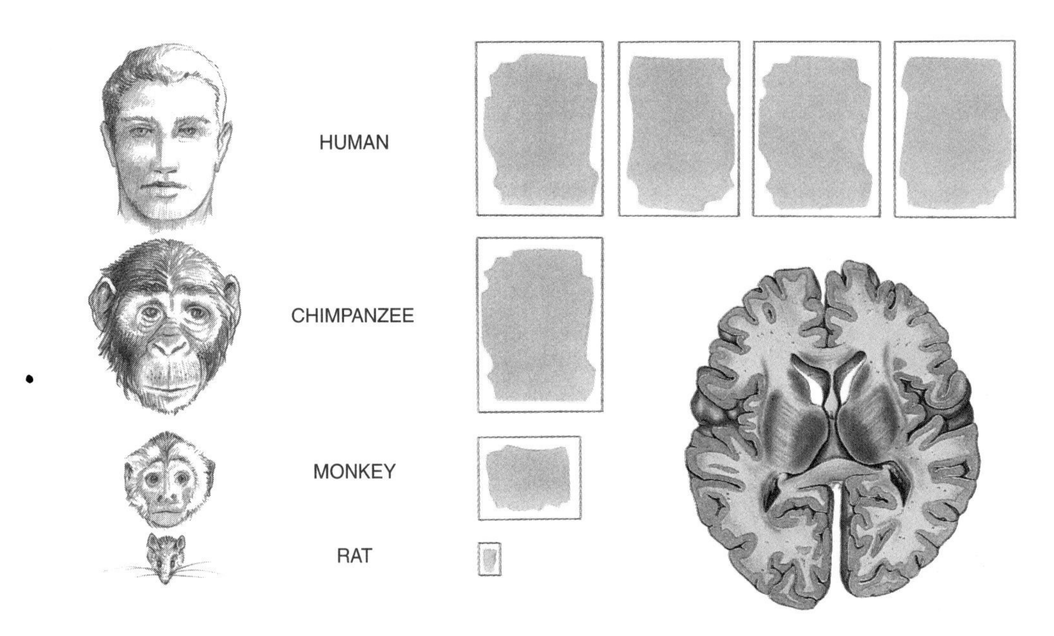

Figure 8.1 CEREBRAL CORTEX is the deeply convoluted surface region of the brain that is most strongly linked to intelligence (*lower right*). A human's cerebral cortex, if flattened, would cover four pages of typing paper; a chimpanzee's would cover only one; a monkey's would cover a postcard; and a rat's would cover a postage stamp.

extensively wrinkled, but were it flattened, it would occupy four sheets of typing paper. A chimpanzee's cortex would fit on one sheet, a monkey's on a postcard, a rat's on a stamp (see Figure 8.1).

Yet a purely quantitative explanation seems incomplete. I will argue that our intelligence arose primarily through the refinement of some brain specialization, such as that for language. The specialization would allow a quantum leap in cleverness and foresight during the evolution of humans from apes. If, as I suspect, that specialization involves a core facility common to language, the planning of hand movements, music and dance, it has even greater explanatory power (see Figure 8.2).

A particularly intelligent person often seems "quick" and capable of juggling many ideas at once. Indeed, the two strongest influences on your IQ score are how many novel questions you can answer in a fixed length of time and how good you are at manipulating half a dozen mental images—as in those analogy questions: A is to B as C is to (D, E or F).

Versatility is another characteristic of intelligence. Most animals are narrow specialists, especially in matters of diet: the mountain gorilla consumes 50 pounds of green leaves each and every day. In comparison, a chimpanzee switches around a lot—it will eat fruit, termites, leaves and even a small monkey or piglet if it is lucky enough to catch one. Omnivores have more basic moves in their general behavior because their ancestors had to switch between many different food sources. They need more sensory templates, too—mental images of things such as foods and predators for which they are "on the lookout." Their behavior

Figure 8.2 BONOBOS and other chimpanzees have a remarkable aptitude for simple language and certain tool-usage skills, such as hammering with stones. Yet compared with those of humans, the abilities of these animals are fairly rudimentary. Human intelligence may have evolved through the enhancement of a core facility that assists with the planning of rapid hand and mouth movements.

emerges through the matching of these sensory templates to responsive movements.

Sometimes animals try out a novel combination of search image and movement during play and find a use for it later. Many animals are playful only as juveniles; being an adult is a serious business (they have all those young mouths to feed). Having a long juvenile period, as apes and humans do, surely aids intelligence. A long life further promotes versatility by affording more opportunities to discover new behaviors.

A social life also gives individuals the chance to mimic the useful discoveries of others. Researchers have seen a troop of monkeys in Japan copy one inventive female's techniques for washing sand off food. Moreover, a social life is full of interpersonal problems to solve, such as those created by pecking orders, that go well beyond the usual environmental challenges to survival and reproduction.

Yet versatility is not always a virtue, and more of it is not always better. As frequent airline travelers know, passengers who have only carry-on bags can get all the available taxicabs while those burdened by three suitcases await their luggage. On the other hand, if the weather is so unpredictable that everyone has to travel with clothing ranging from swimsuits to Arctic parkas, the "jack of all trades" has an advantage over the "master" of one. And so it is with behavioral versatility and brain size.

When chimpanzees in Uganda arrive at a grove of fruit trees, they often discover that the efficient local monkeys are already speedily stripping the trees of edible fruit. The chimps can turn to termite fishing or perhaps catch a monkey and eat it, but in practice their population is severely limited by that competition, despite a brain twice the size of their specialist rivals'.

Whether versatility is advantageous depends on the timescales: for both the modern traveler and the evolving ape, it is how fast the weather changes and how long the trip lasts. Paleoclimatologists have discovered that many parts of the earth suffer sudden climate changes, as abrupt in onset as a decade-long drought but lasting for centuries (see Figure 8.3). A climatic flip that eliminated fruit trees would be disastrous for many monkey species. It would hurt the more omnivorous animals, too, but they could make do with other foods, and eventually they would enjoy a population boom when the food crunch ended and few of their competitors remained.

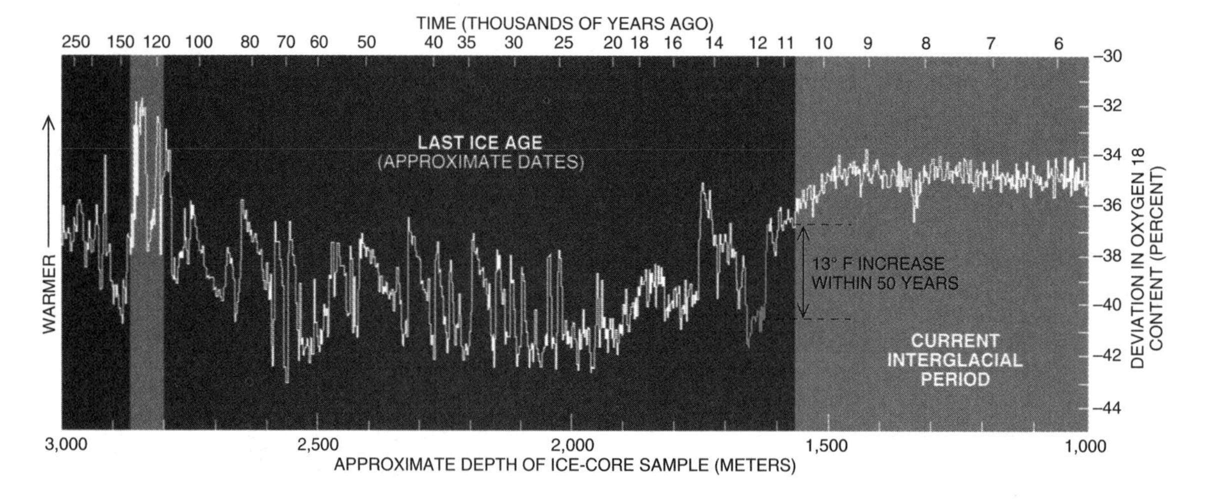

Figure 8.3 RAPID CLIMATE CHANGES may have promoted behavioral flexibility in the ancestors of modern humans. During the last ice age, the average temperature was much lower than it is today, but it was also subject to large, abrupt fluctuations that sometimes lasted for centuries. During one climatic oscillation, for example (*red line*), the temperature rose 13 degrees Fahrenheit, rainfall increased by 50 percent and the severity of dust storms fell, all in the space of a few decades. Cold periods began just as suddenly. Early humans may have needed greater intellectual resources to survive these changes. (This graph is based on work by W. Dansgaard of the University of Copenhagen and his colleagues using Greenland ice cores.)

Although Africa was cooling and drying as upright posture was becoming established four million years ago, brain size did not change much. The fourfold expansion of the hominid brain did not start until the ice ages began, 2.5 million years ago. Ice cores from Greenland show frequent abrupt cooling episodes superimposed on the more stately rhythms of ice advance and retreat. Entire forests disappeared within several decades because of drastic drops in temperature and rainfall. The warm rains returned with equal suddenness several centuries later.

The evolution of anatomical adaptations in the hominids could not have kept pace with these abrupt climate changes, which would have occurred within the lifetime of single individuals. Still, these environmental fluctuations could have promoted the incremental accumulation of mental abilities that conferred greater behavioral flexibility.

One of the additions during the ice ages was the capacity for human language. In most of us, the brain area critical to language is located just above our left ear. Monkeys lack this left lateral language area: their vocalizations (and simple emotional utterances in humans) employ a more primitive language area near the corpus callosum, the band of fibers connecting the cerebral hemispheres.

Language is the most defining feature of human intelligence: without syntax—the orderly arrangement of verbal ideas—we would be little more clever than a chimpanzee. For a glimpse of life without syntax, we can look to the case of Joseph, an 11-year-old deaf boy. Because he could not hear spoken language and had never been exposed to fluent sign language, Joseph did not have the chance to learn syntax during the critical years of early childhood.

As neurologist Oliver W. Sacks described him in *Seeing Voices:* "Joseph saw, distinguished, categorized, used; he had no problems with *perceptual* categorization or generalization, but he could not, it seemed, go much beyond this, hold abstract ideas in mind, reflect, play, plan. He seemed completely literal—unable to juggle images or hypotheses or possibilities, unable to enter an imaginative or figurative realm. . . . He seemed, like an animal, or an infant, to be stuck in the present, to be confined to literal and immediate perception, though made aware of this by a consciousness that no infant could have."

To understand why humans are so intelligent, we need to understand how our ancestors remodeled the apes' symbolic repertoire and enhanced it by inventing syntax. Wild chimpanzees use about three dozen different vocalizations to convey about three dozen different meanings. They may repeat a sound to intensify its meaning, but they do not string together three sounds to add a new word to their vocabulary.

We humans also use about three dozen vocalizations, called phonemes. Yet only their combinations have content: we string together meaningless sounds to make meaningful words. No one has yet explained how our ancestors got over the hump of replacing "one sound/one meaning" with a sequential combinatorial system of meaningless phonemes, but it is probably one of the most important advances that took place during ape-to-human evolution.

Furthermore, human language uses strings of strings, such as the word phrases that make up this sentence. The simplest ways of generating word collections, such as pidgin dialects (or my tourist German), are known as protolanguage. In a protolanguage, the association of the words carries the message, with perhaps some assistance from customary word order (such as the subject-verb-object order in English declarative sentences).

Our closest animal cousins, the common chimpanzee and the bonobo (pygmy chimpanzee), can achieve surprising levels of language comprehension when motivated by skilled teachers. Kanzi, the most accomplished bonobo, can interpret sentences he has never heard before, such as "Go to the office and bring back the red ball," about as well as a 2.5-year-old child. Neither Kanzi nor the child constructs such sentences independently, but each can demonstrate understanding (see Figure 8.4).

With a year's experience in comprehension, the child starts constructing fancier sentences. The rhyme about the house that Jack built ("This is the farmer sowing the corn/That kept the cock that crowed in the morn/. . . That lay in the house that Jack built") is an extreme case of nesting word phrases inside one another, yet even preschoolers understand how "that" changes its meaning.

Syntax has treelike rules of reference that enable us to communicate quickly—sometimes with fewer than 100 sounds strung together—who did what to whom, where, when, why and how. Generating and speaking a unique sentence demonstrate whether you know the rules of syntax well enough to avoid ambiguities. Even children of low intelligence acquire syntax effortlessly by listening,

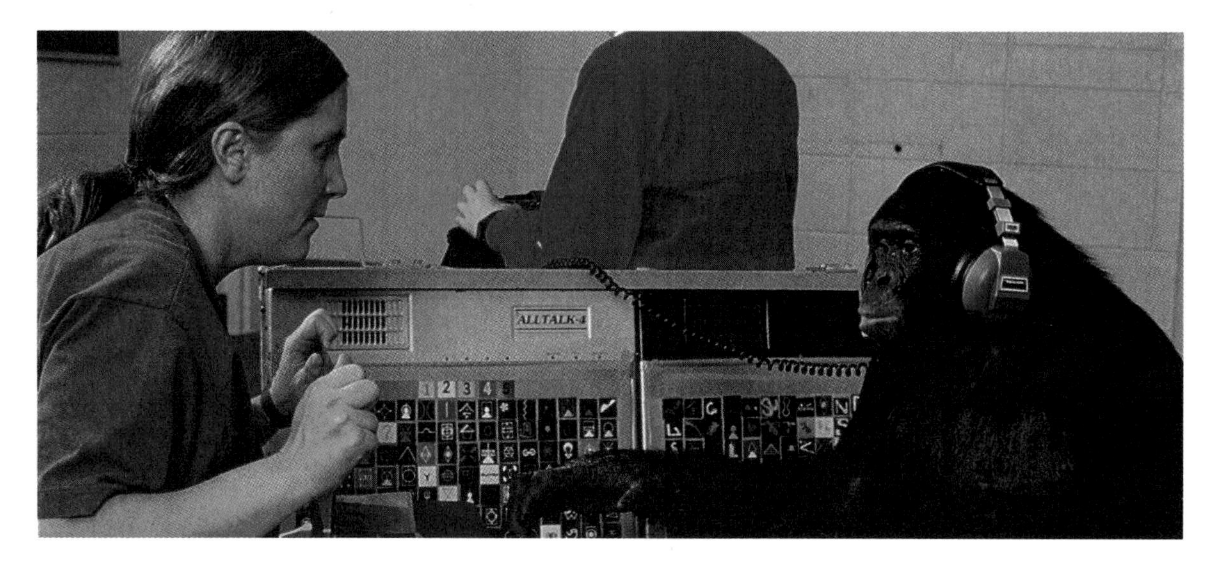

Figure 8.4 KANZI, a bonobo, has been reared at Georgia State University in a language-using environment. By pointing at symbols that represent various words, Kanzi can construct requests much like those of a two-year-old child. His comprehension is as good as that of a 2.5-year-old. Language experiments on bonobos ask how much of syntax is uniquely human.

although intelligent deaf children like Joseph may miss out (see Figure 8.5).

Something close to verbal syntax also seems to contribute to another outstanding feature of human intelligence, the ability to plan. Aside from hormonally triggered preparations for winter and mating, animals exhibit surprisingly little evidence of planning more than a few minutes ahead. Some chimpanzees use long twigs to pull termites from their nests, yet as Jacob Bronowski observed, none of the termite-fishing chimps "spends the evening going round and tearing off a nice tidy supply of a dozen probes for tomorrow."

Short-term planning does occur to an extent, and it seems to allow a crucial increment in social intelligence. Deception is seen in apes, but seldom in monkeys. A chimp may give a call signaling that she has found food at one location, then quietly circle back through the dense forest to where she actually found the food. While the other chimps beat the bushes at the site of the food cry, she eats without sharing.

The most difficult responses to plan are those to unique situations. They require imagining multiple scenarios, as when a hunter plots various approaches to a deer or a futurist spins three scenarios bracketing what an industry will look like in another decade. Compared with apes, humans do a lot of that—we can heed the admonition sometimes attributed to British statesman Edmund Burke: "The public interest requires doing today those things that men of intelligence and goodwill would wish, five or 10 years hence, had been done."

Human planning abilities may stem from our talent for building syntactical, string-based conceptual structures larger than sentences. As the writer Kathryn Morton observes about narrative:

> *The first sign that a baby is going to be a human being and not a noisy pet comes when he begins naming the world and demanding the stories that connect its parts. Once he knows the first of these he will instruct his teddy bear, enforce his world-view on victims in the sandlot, tell himself stories of what he is doing as he plays and forecast stories of what he will do when he grows up. He will keep track of the actions of others and relate deviations to the person in charge. He will want a story at bedtime.*

Our abilities to plan gradually develop from childhood narratives and are a major foundation for ethical choices, as we imagine a course of action, imagine its effects on others and decide whether or not to do it.

In this way, syntax raises intelligence to a new level. By borrowing the mental structures for syntax to judge other combinations of possible actions, we can extend our planning abilities and our intelligence. To some extent, we do this by talking silently to ourselves, making narratives out of what might happen next and then applying syntax-like rules of combination to rate a scenario as dangerous nonsense, mere nonsense, possible, likely or logical. But our thinking is not limited to language-like constructs. Indeed, we may shout, "Eureka!" when feeling a set of mental relationships click into place and yet have trouble expressing them verbally.

Language and intelligence are so powerful that we might think evolution would naturally favor their increase. As evolutionary theorists are fond of demonstrating, however, the fossil record is full of plateaus. Evolution often follows indirect routes rather than "progressing" through adaptations. To account for the breadth of our abilities, we need to look at improvements in common-core facilities. Environments that give the musically gifted an evolutionary advantage over the tone deaf are difficult to imagine, but there are multifunctional brain mechanisms whose improvement for one critical function might incidentally aid other functions.

We humans certainly have a passion for stringing things together: words into sentences, notes into melodies, steps into dances, narratives into games with rules of procedure. Might stringing things together be a core facility of the brain, one commonly useful to language, storytelling, planning, games and ethics? If so, natural selection for any of these talents might augment their shared neural machinery, so that an improved knack for syntactical sentences would automatically expand planning abilities, too. Such carryover is what Charles Darwin called functional change in anatomical continuity, distinguishing it from gradual adaptation. To some extent, music and dance are surely secondary uses of neural machinery shaped by sequential behaviors more exposed to natural selection, such as language.

As improbable as the idea initially seems, the brain's planning of ballistic movements may have once promoted language, music and intelligence. Ballistic movements are extremely rapid actions of the limbs that, once initiated, cannot be modified. Striking a nail with a hammer is an example. Apes have only elementary forms of the ballistic arm movements at which humans are expert—hammering, clubbing and throwing. These movements are integral to the manufacture and use of tools and hunting weapons: in some settings, hunting and

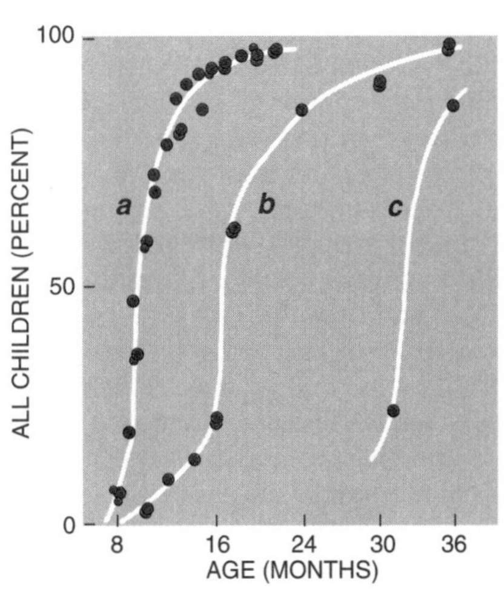

Figure 8.5 ACQUISITION OF LANGUAGE by children occurs quickly and naturally through exposure to adults.

Figure 8.6 THROWING is a ballistic movement at which humans excel, despite the lack of effective feedback from the arm during most of the throw. Before a pitch starts, the brain must plan the sequence of muscle contractions that will launch the ball toward a target. Some of the neural mechanisms that plan such movements may also facilitate other types of planning.

toolmaking were probably important additions to hominids' basic survival strategies.

Ballistic movements require a surprising amount of planning (see Figure 8.6). Slow movements leave time for improvisation: when raising a cup to your lips, if the cup is lighter than you remembered, you can correct its trajectory before it hits your nose. Thus, a complete plan is not needed. You start in the right general direction and then correct your path, just as a moon rocket does.

For sudden limb movements lasting less than one fifth of a second, feedback corrections are largely ineffective because reaction times are too long. The brain has to determine every detail of the movement in advance, as though it were silently punching a roll of music for a player piano.

Hammering requires scheduling the exact sequence of activation for dozens of muscles. The problem of throwing is further compounded by the launch window—the range of times in which a projectile can be released to hit a target. When the distance to a target doubles, the launch window becomes eight times narrower; statistical arguments indicate that programming a reliable throw would then require the activity of 64 times as many neurons.

If mouth movements rely on the same core facility for sequencing that ballistic hand movements do, then enhancements in language skills might improve dexterity, and vice versa. Accurate throwing abilities open up the possibility of eating meat regularly, of being able to survive winter in a temperate zone. The gift of speech would be an incidental benefit—a free lunch, as it were, because of the linkage.

Is there actually a sequencer common to movement and language? Much of the brain's coordination of movement occurs at a subcortical level in the basal ganglia or the cerebellum, but novel combinations of movements tend to depend on the premotor and prefrontal cortex. Two major lines of evidence point to cortical specialization for sequencing, and both suggest that the lateral language area has a lot to do with it.

Doreen Kimura of the University of Western Ontario [see "Sex Differences in the Brain"; SCIENTIFIC AMERICAN, September 1992] has found that stroke patients with language problems (aphasia) resulting from damage to left lateral brain areas also have considerable difficulty executing unfamiliar sequences of hand and arm movements (apraxia). By electrically stimulating the brains of patients being operated on for epilepsy, George A. Ojemann of the University of Washington has also shown that at the center of the left lateral areas specialized for language lies a region involved in listening to sound sequences (see Figure 8.7). This perisylvian region seems equally involved in producing oral-facial movement sequences—even nonlanguage ones.

These discoveries reveal that parts of the "language cortex," as people sometimes think of it, serve a far more generalized function than had been suspected. The language cortex is concerned with novel sequences of various kinds: both sensations and movements, for both the hands and the mouth.

The big problem with inventing sequences and producing original behaviors is safety. Even simple reversals in order can be dangerous, as in "Look *after* you leap." Our capacity to make analogies and mental models gives us a measure of protection, however. We humans can simulate future courses of action and weed out the nonsense off-line; as philosopher Karl Popper said, this "permits our hypotheses to die in our stead." Creativity—indeed, the whole high end of intelligence and consciousness—involves playing mental games that shape up quality before acting. What kind of mental machinery might it take to do something like that?

By 1874, just 15 years after Darwin published *The Origin of Species*, the American psychologist William James was talking about mental processes operating in a Darwinian manner. In effect, he suggested, ideas might somehow "compete" with one

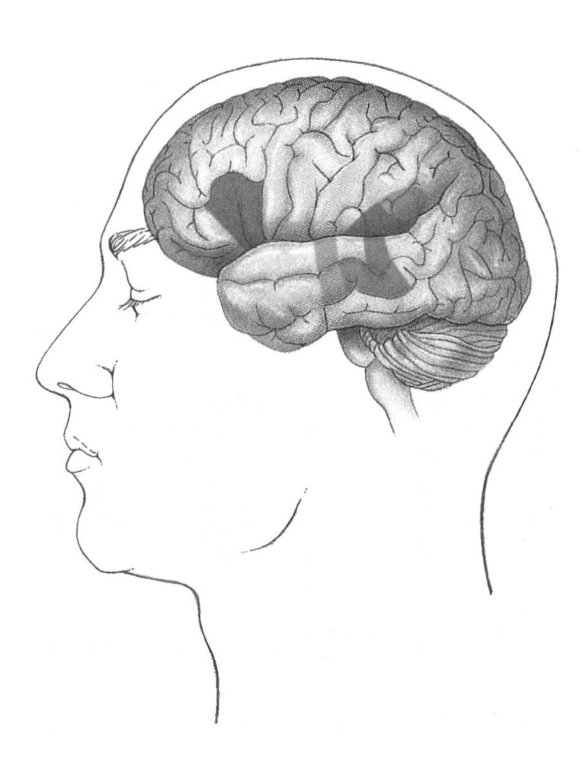

Figure 8.7 SPECIALIZED SEQUENCING REGION of the left cerebral cortex is involved both in listening to spoken language and in producing oral-facial movements. The shading (from the data of George A. Ojemann of the University of Washington) reflects the relative involvement in these activities.

another in the brain, leaving only the best or "fittest." Just as Darwinian evolution shaped a better brain in two million years, a similar Darwinian process operating within the brain might shape intelligent solutions to problems on the timescale of thought and action (see box on "A Meditation on Creative Thought").

Researchers have demonstrated that a Darwinian process operating on an intermediate timescale of days governs the immune response following a vaccination. Through a series of cellular generations spanning several weeks, the immune system produces defensive antibody molecules that are better and better "fits" against invaders. By abstracting the essential features of a Darwinian process from what is known about species evolution and immune responses, we can see that any "Darwin machine" must have six properties.

First, it must operate on patterns of some type; in genetics, they are strings of DNA bases, but patterns of brain activity associated with a thought might qualify. Second, copies are made of these patterns. (Indeed, that which is reliably copied defines a unit pattern.) Third, patterns must occasionally vary, whether through mutations, copying errors or a reshuffling of their parts.

Fourth, variant patterns must compete to occupy some limited space (as when bluegrass and crabgrass compete for my backyard). Fifth, the relative reproductive success of the variants is influenced by their environment; this result is what Darwin called natural selection. And, finally, the makeup of the next generation of patterns depends on which variants survive to be copied. The patterns of the next generation will be variations spread around the currently successful ones. Many of these new variants will be less successful than their parents, but some may be more so.

Sex and climatic change may not be numbered among the six essentials, but they add spice and speed to a Darwinian process, whether it operates in milliseconds or millennia. Note that an "essential" is not Darwinian by itself: for example, selective survival can be seen when flowing water carries away sand and leaves pebbles behind.

Let us consider how these principles might apply to the evolution of an intelligent guess inside the brain. Thoughts are combinations of sensations and memories—in a way, they are movements that have not happened yet (and maybe never will). They exist as patterns of spatiotemporal activity in the brain, each representing an object, action or abstraction. I estimate that a single cerebral code minimally involves a few hundred cortical neurons within a millimeter of one another either firing or keeping quiet.

Evoking a memory is simply a matter of reconstituting such an activity pattern, according to psychologist Donald O. Hebb's cell-assembly hypothesis [see "The Mind and Donald O. Hebb," by Peter M. Milner; SCIENTIFIC AMERICAN, January

A Meditation on Creative Thought

I believe the brain plays a game—some parts providing the stimuli, the others the reactions, and so on. . . . One is only consciously aware of something in the brain which acts as a summarizer or totalizer of the process going on and that probably consists of many parts acting simultaneously on each other. Clearly only the one-dimensional chain of syllogisms which constitutes thinking can be communicated verbally or written down. . . . If, on the other hand, I want to do something new or original, then it is no longer a question of syllogism chains. When I was a boy I felt that the role of rhyme in poetry was to compel one to find the unobvious because of the necessity of finding a word which rhymes. This forces novel associations and almost guarantees deviations from routine chains or trains of thought. It becomes paradoxically a sort of automatic mechanism of originality. . . . And what we call talent or perhaps genius itself depends to a large extent on the ability to use one's memory properly to find the analogies . . . [which] are essential to the development of new ideas.

—Stanislaw M. Ulam,
Adventures of a Mathematician, 1976

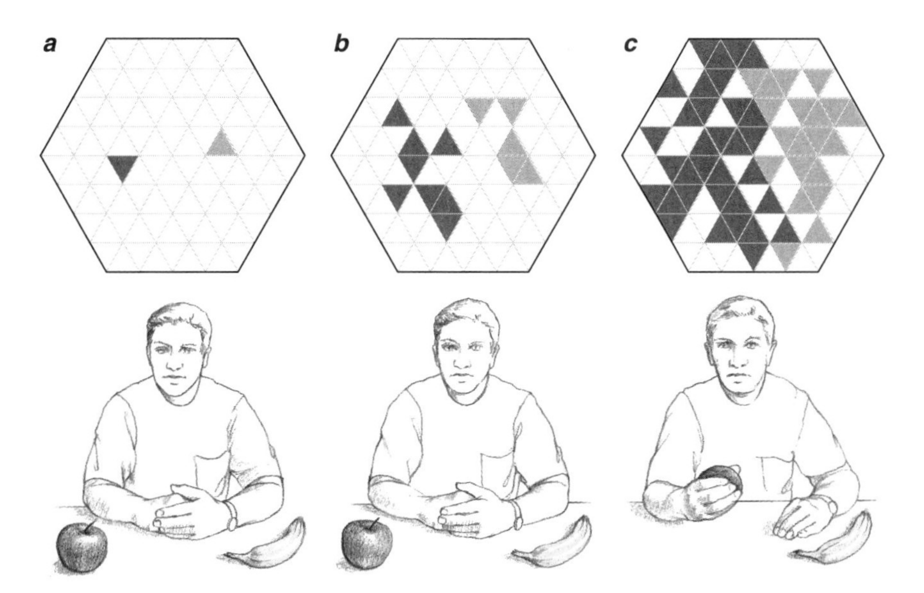

Figure 8.8 DARWINIAN MODEL OF THINKING suggests that ideas compete for "workspace" within the brain. When a person is choosing between an apple and a banana (*a*), spatiotemporal patterns of neural activity representing these possibilities (*red for apple, yellow for banana*) may appear in the cortex (*hexagon*). Copies of each pattern proliferate at different rates, depending on the individual's experiences and sensory impressions (*b*). Eventually, the number of copies of one pattern passes a threshold, and the person makes that choice—in this case, to take the apple (*c*).

1993]. Long-term memories are frozen patterns waiting for signals of near resonance to reawaken them, like ruts in a washboarded road waiting for a passing car to re-create a bouncing spatiotemporal pattern.

Some "cerebral ruts" are permanent, whereas others are short-lived. Short-term memories are just temporary alterations in the strengths of synaptic connections between neurons, left behind by the last spatiotemporal pattern to occupy a patch of cortex; this "long-term potentiation" may fade in a matter of minutes. The transition from short- to long-term patterning is not well understood, but structural alterations may sometimes follow potentiation, such that the synaptic connections between neurons are made strong and permanent, hardwiring the pattern of neural activity into the brain.

A Darwinian model of mind suggests that an activated memory can compete with others for "workspace" in the cortex. Perceptions of the thinker's current environment and memories of past environments may bias that competition and shape an emerging thought (see Figure 8.8).

An active cerebral code moves from one part of the brain to another by making a copy of itself, much as a facsimile machine re-creates a copy of a pattern on a distant sheet of paper. The cerebral cortex also has circuitry for copying spatiotemporal patterns in an immediately adjacent region less than a millimeter away, although our present imaging techniques lack enough resolution to see the copying in progress. Repeated copying of the minimal pattern could colonize a region, rather the way a crystal grows or wallpaper repeats an elementary pattern.

The picture that emerges from these theoretical considerations is one of a quilt, some patches of which enlarge at the expense of their neighbors as one code copies more successfully than another. As you try to decide whether to pick an apple or a banana from the fruit bowl, so my theory goes, the cerebral code for "apple" may be having a cloning competition with the one for "banana." When one code has enough active copies to trip the action circuits, you might reach for the apple.

But the banana codes need not vanish: they could linger in the background as subconscious thoughts and undergo variations. When you try to remember someone's name, initially without success, the candidate codes might continue copying for the next half an hour until, suddenly, Jane Smith's name seems to "pop into your mind" because your variations on the spatiotemporal theme finally hit a resonance and create a critical mass of identical copies. Our conscious thought may be only the currently dominant pattern in the copying competition, with many other variants competing for dominance, one of which will win a moment later when your thoughts seem to shift focus.

It may be that Darwinian processes are only the frosting on the cognitive cake, that much of our thinking is routine or bound by rules. But we often deal with novel situations in creative ways, as when you decide what to fix for dinner tonight. You survey what is already in the refrigerator and on the kitchen shelves. You think about a few alternatives, keeping track of what else you might have to fetch from the grocery store. All this can flash through your mind within seconds—and that is probably a Darwinian process at work.

In phylogeny and its ontogeny, human intelligence first solves movement problems and only later graduates to ponder more abstract ones. An artificial or extraterrestrial intelligence freed of the necessity of finding food and avoiding predators might not need to move—and so might lack the what-happens-next orientation of human intelligence. There may be other ways in which high intelligence can be achieved, but up-from-movement is the known paradigm.

It is difficult to estimate how often high intelligence might emerge, given how little we know about the demands of long-term species survival and the courses evolution can follow. We can, however, compare the prospects of different species by asking how many elements of intelligence each has amassed.

Does the species have a wide repertoire of movements, concepts or other tools? Does it have tolerance for creative confusion that allows individuals to invent categories occasionally? (Primatologist Duane M. Rumbaugh of Georgia State University has noted that small monkeys and prosimians, such as lemurs, often get trapped into repeating the first set of discrimination rules they are taught, unlike the more advanced rhesus monkeys and apes.)

Does each individual have more than half a dozen mental workspaces for concurrently holding different concepts? Does it have so many that it loses our human tendency to "chunk" certain concepts, as when we create the word "ambivalence" to stand for a whole sentence's worth of description? Can individuals establish new relations between the concepts in their workspaces? These relations should be fancier than "is a" and "is larger than," which many animals can grasp. Treelike relations seem particularly important for linguistic structures; our ability to compare two relations (analogy) enables operations in a metaphorical space.

Can individuals mold and refine their ideas offline, before acting in the real world? Does that process involve all six of the essential Darwinian features, as well as some accelerating factors—shortcuts that allow the process to start from something more than a primitive level? Can individuals make guesses about both long-term strategies and short-term tactics, so that they can make moves that will advantageously set the stage for future feats?

Chimps and bonobos may be missing a few of these elements, but they are doing better than the present generation of artificial-intelligence programs. Even in entities with all the elements, we would expect considerable variation in intelligence because of individual differences in processing speed, in perseverance, in implementing shortcuts and in finding the appropriate level of abstraction when using analogies.

Why are there not more species with such complex mental states? A little intelligence can be a dangerous thing. A beyond-the-apes intelligence must constantly navigate between the twin hazards of dangerous innovation and a conservatism that ignores what the Red Queen explained to Alice in *Through the Looking Glass*: " . . . it takes all the running you can do, to keep in the same place." Foresight is our special form of running, essential for the intelligent stewardship that Stephen Jay Gould of Harvard University warns is needed for longer-term survival: "We have become, by the power of a glorious evolutionary accident called intelligence, the stewards of life's continuity on earth. We did not ask for this role, but we cannot abjure it. We may not be suited for it, but here we are."

Will Robots Inherit the Earth?

Yes, as we engineer replacement bodies and brains using nanotechnology. We will then live longer, possess greater wisdom and enjoy capabilities as yet unimagined.

• • •

Marvin Minsky

Early to bed and early to rise,
Makes a man healthy, wealthy, and wise.
—Benjamin Franklin

Everyone wants wisdom and wealth. Nevertheless, our health often gives out before we achieve them. To lengthen our lives and improve our minds, we will need to change our bodies and brains. To that end, we first must consider how traditional Darwinian evolution brought us to where we are. Then we must imagine ways in which novel replacements for worn body parts might solve our problems of failing health. Next we must invent strategies to augment our brains and gain greater wisdom. Eventually, using nanotechnology, we will entirely replace our brains. Once delivered from the limitations of biology, we will decide the length of our lives—with the option of immortality—and choose among other, unimagined capabilities as well.

In such a future, attaining wealth will be easy; the trouble will be in controlling it. Obviously, such changes are difficult to envision, and many thinkers still argue that these advances are impossible, particularly in the domain of artificial intelligence. But the sciences needed to enact this transition are already in the making, and it is time to consider what this new world will be like.

Such a future cannot be realized through biology. In recent times we have learned much about health and how to maintain it. We have devised thousands of specific treatments for specific diseases and disabilities. Yet we do not seem to have increased the maximum length of our life span. Benjamin Franklin lived for 84 years, and except in popular legends and myths no one has ever lived twice that long. According to the estimates of Roy L. Walford, professor of pathology at the University of California at Los Angeles School of Medicine, the average human lifetime was about 22 years in ancient Rome, was about 50 in the developed countries in 1900, and today stands at about 75 in the U.S. Despite this increase, each of those curves seems to terminate sharply near 115 years. Centuries of improvements in health care have had no effect on that maximum (see Figure 9.1).

Why are our life spans so limited? The answer is simple: natural selection favors the genes of those with the most descendants. Those numbers tend to grow exponentially with the number of generations, and so natural selection prefers the genes of those who reproduce at earlier ages. Evolution

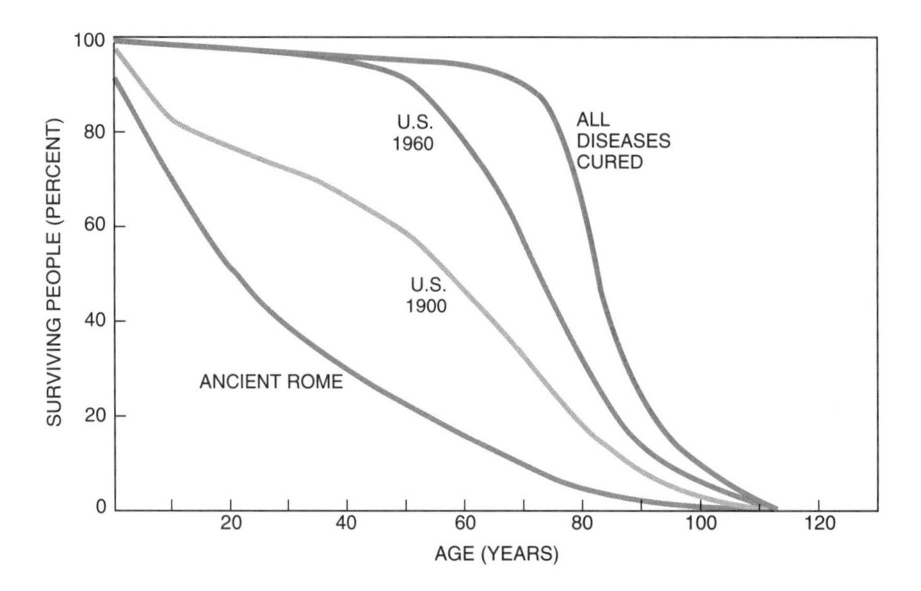

Figure 9.1 HUMAN LIFE SPAN has increased on average over time as economic conditions have improved. In ancient Rome (*brown*) the average lifetime was 22 years, in developed countries around 1900 (*blue*) it was 50, and now in the U.S. (*dark blue*) it stands at 75. Still, these curves share the same maximum. Even if we found cures for every plague (*red*), our bodies would probably wear out after roughly 115 years.

does not usually preserve genes that lengthen lives beyond that amount adults need to care for their young. Indeed, it may even favor offspring who do not have to compete with living parents. Such competition could promote the accretion of genes that cause death. For example, after spawning, the Mediterranean octopus promptly stops eating and starves itself. If a certain gland is removed, the octopus continues to eat and lives twice as long. Many other animals are programmed to die soon after they cease reproducing. Exceptions to this phenomenon include animals such as ourselves and elephants, whose progeny learn a great deal from the social transmission of accumulated knowledge.

We humans appear to be the longest-lived warm-blooded animals. What selective pressure might have led to our present longevity, which is almost twice that of our other primate relatives? The answer is related to wisdom. Among all mammals our infants are the most poorly equipped to survive by themselves. Perhaps we need not only parents but grandparents, too, to care for us and to pass on precious survival tips.

Even with such advice there are many causes of mortality to which we might succumb. Some deaths result from infections. Our immune systems have evolved versatile ways to cope with most such diseases. Unhappily, those very same immune systems often injure us by treating various parts of ourselves as though they, too, were infectious invaders. This autoimmune blindness leads to diseases such as diabetes, multiple sclerosis, rheumatoid arthritis and many others.

We are also subject to injuries that our bodies cannot repair: accidents, dietary imbalances, chemical poisons, heat, radiation and sundry other influences can deform or chemically alter the molecules of our cells so that they are unable to function. Some of these errors get corrected by replacing defective molecules. Nevertheless, when the replacement rate is too low, errors build up. For example, when the proteins of the eyes' lenses lose their elasticity, we lose our ability to focus and need bifocal spectacles—a Franklin invention.

The major natural causes of death stem from the effects of inherited genes. These genes include those that seem to be largely responsible for heart disease and cancer, the two biggest causes of mortality, as well as countless other disorders, such as cystic fibrosis and sickle cell anemia. New technologies should be able to prevent some of these disorders by replacing those genes.

Most likely, senescence is inevitable in all biological organisms. To be sure, certain species (including some varieties of fish, tortoises and lobsters) do not appear to show any systematic increase in mortality as they age. These animals seem to die mainly from external causes, such as predators or starvation. All the same, we have no records of animals that have lived for as long as 200 years—although this lack does not prove that none exist (see Figure 9.2). Walford and many others believe a carefully designed diet, one seriously restricted in calories, can significantly increase a human's life span but cannot ultimately prevent death.

By learning more about our genes, we should be able to correct or at least postpone many conditions that still plague our later years. Yet even if we found a cure for each specific disease, we would still have to face the general problem of "wearing out." The normal function of every cell involves thousands of chemical processes, each of which sometimes makes random mistakes. Our bodies use many kinds of correction techniques, each triggered by a specific type of mistake. But those random errors happen in so many different ways that no low-level scheme can correct them all.

The problem is that our genetic systems were not designed for very long term maintenance. The relation between genes and cells is exceedingly indirect; there are no blueprints or maps to guide our genes as they build or rebuild the body. To repair defects on larger scales, a body would need some kind of catalogue that specified which types of cells should be located where. In computer programs it is easy to install such redundancy. Many computers maintain unused copies of their most critical system programs and routinely check their integrity. No animals have evolved similar schemes, presumably because such algorithms cannot develop through natural selection. The trouble is that error correction would stop mutation, which would ultimately slow the rate of evolution of an animal's descendants so much that they would be unable to adapt to changes in their environments.

Figure 9.2 ANTS from the species *Lasius niger* are shown here swarming. An *L. niger* queen ant is known to have lived for 27 years. No sexually reproducing animal on record has lived more than 200 years, but some may exist. Although certain species seem to die only from external causes, such as predation or starvation, senescence is probably inevitable in all biological organisms.

Could we live for several centuries simply by changing some number of genes? After all, we now differ from our relatives, the gorillas and chimpanzees, by only a few thousand genes—and yet we live almost twice as long. If we assume that only a small fraction of those new genes caused that increase in life span, then perhaps no more than 100 or so of those genes were involved. Even if this turned out to be true, though, it would not guarantee that we could gain another century by changing another 100 genes. We might need to change just a few of them—or we might have to change a good many more.

Making new genes and installing them are slowly becoming feasible. But we are already exploiting another approach to combat biological wear and tear: replacing each organ that threatens to fail with a biological or artificial substitute. Some replacements are already routine. Others are on the horizon (see Figure 9.3). Hearts are merely clever pumps. Muscles and bones are motors and beams. Digestive systems are chemical reactors. Eventually, we will find ways to transplant or replace all these parts.

But when it comes to the brain, a transplant will not work. You cannot simply exchange your brain for another and remain the same person. You would lose the knowledge and the processes that constitute your identity. Nevertheless, we might be able to replace certain worn-out parts of brains by transplanting tissue-cultured fetal cells. This procedure would not restore lost knowledge, but that might not matter as much as it seems. We probably store each fragment of knowledge in several different places, in different forms. New parts of the brain could be retrained and reintegrated with the rest, and some of that might even happen spontaneously.

Even before our bodies wear out, I suspect that we often run into limitations in our brains' abilities. As a species, we seem to have reached a plateau in our intellectual development. There is no sign that we are getting smarter. Was Albert Einstein a better scientist than Isaac Newton or Archimedes? Has any playwright in recent years topped William Shakespeare or Euripides? We have learned a lot in 2,000 years, yet much ancient wisdom still seems sound, which makes me think we have not been making much progress. We still do not know how to resolve conflicts between individual goals and global interests. We are so bad at making important decisions that, whenever we can, we leave to chance what we are unsure about.

Why is our wisdom so limited? Is it because we do not have the time to learn very much or that we lack enough capacity? Is it because, according to popular accounts, we use only a fraction of our brains? Could better education help? Of course, but only to a point. Even our best prodigies learn no more than twice as quickly as the rest. Everything takes us too long to learn because our brains are so terribly slow. It would certainly help to have more time, but longevity is not enough. The brain, like other finite things, must reach some limits to what it can learn. We do not know what those limits are; perhaps our brains could keep learning for several more centuries. But at some point, we will need to increase their capacity.

The more we learn about our brains, the more ways we will find to improve them. Each brain has hundreds of specialized regions. We know only a little about what each one does or how it does it, but as soon as we find out how any one part works, researchers will try to devise ways to extend that part's capacity. They will also conceive of entirely new abilities that biology has never provided. As these inventions grow ever more prevalent, we will try to connect them to our brains, perhaps through millions of microscopic electrodes inserted into the great nerve bundle called the corpus callosum, the largest databus in the brain. With further advances, no part of the brain will be out-of-bounds for attaching new accessories. In the end, we will find ways to replace every part of the body and brain and thus repair all the defects and injuries that make our lives so brief (see Figure 9.4).

Needless to say, in doing so we will be making ourselves into machines. Does this mean that machines will replace us? I do not feel that it makes much sense to think in terms of "us" and "them." I much prefer the attitude of Hans P. Moravec of

Figure 9.3 COG, under construction at the Massachusetts Institute of Technology, will have mechanical eyes, ears and arms wired to a network of microprocessors that act as its brain. Cog's creators hope that by interacting with its environment the system will learn to recognize faces, track objects and otherwise respond to a host of visual and auditory stimuli, as would an infant. If the project succeeds, Cog will be the most sophisticated robot assembled to date.

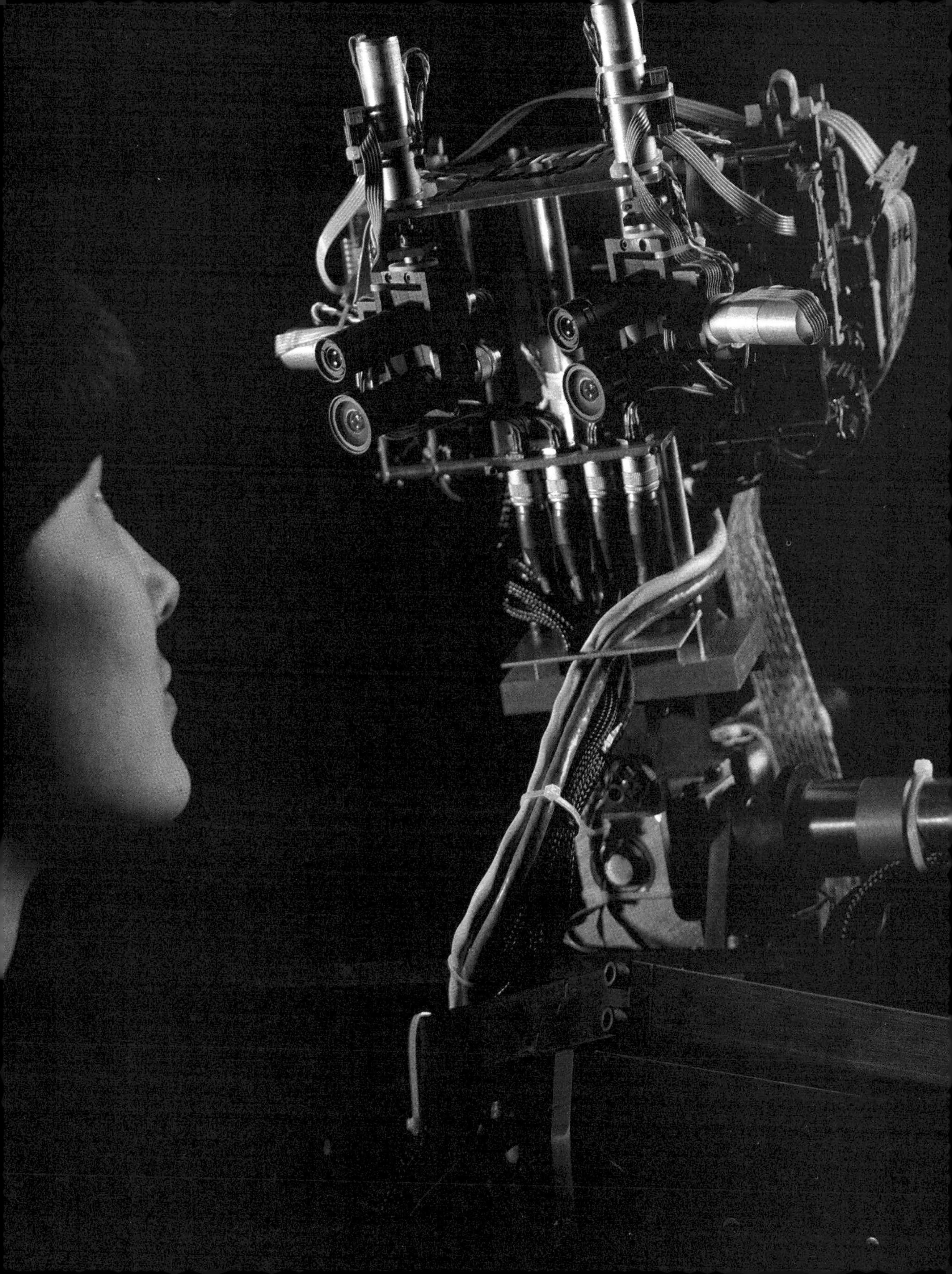

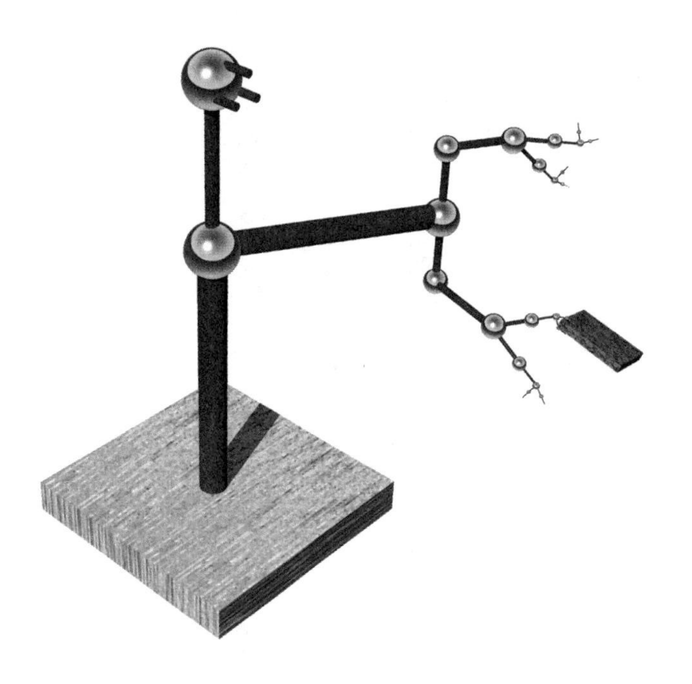

Figure 9.4 ROBOTIC TREE-HAND, designed (but not yet built) by the author and, independently, by Hans P. Moravec of Carnegie Mellon University, is composed of many similar units of different sizes. Because of this uniformity, such robots should be easy to build in the future. At each smaller scale, a tree robot has twice as many units—a pattern not unlike that of the human frame.

Carnegie Mellon University, who suggests that we think of these future intelligent machines as our own "mind-children."

In the past we have tended to see ourselves as a final product of evolution, but our evolution has not ceased. Indeed, we are now evolving more rapidly, though not in the familiar, slow Darwinian way. It is time that we started to think about our new emerging identities. We can begin to design systems based on inventive kinds of "unnatural selection" that can advance explicit plans and goals and can also exploit the inheritance of acquired characteristics. It took a century for evolutionists to train themselves to avoid such ideas—biologists call them "teleological" and "Lamarckian"—but now we may have to change those rules.

Almost all the knowledge we amass is embodied in various networks inside our brains. These networks consist of huge numbers of tiny nerve cells and smaller structures, called synapses, that control how signals jump from one nerve cell to another. To make a replacement of a human brain, we would need to know something about how each of the synapses relates to the two cells it joins. We would also have to know how each of those structures responds to the various electric fields, hormones, neurotransmitters, nutrients and other chemicals that are active in its neighborhood. A human brain contains trillions of synapses, so this is no small requirement.

Fortunately, we would not need to know every minute detail. If details were important, our brains would not work in the first place. In biological organisms, each system has generally evolved to be insensitive to most of what goes on in the smaller subsystems on which it depends. Therefore, to copy a functional brain it should suffice to replicate just enough of the function of each part to produce its important effects on other parts.

Suppose we wanted to copy a machine, such as a brain, that contained a trillion components. Today we could not do such a thing (even with the necessary knowledge) if we had to build each component separately. But if we had a million construction machines that could each build 1,000 parts per second, our task would take mere minutes (see Figure 9.5). In the decades to come, new fabrication machines will make this possible. Most present-day manufacturing is based on shaping bulk materials. In contrast, nanotech-

Figure 9.5 MICROMOTOR is shown here below a pinpoint. As ways are found to make even smaller devices, nanotechnologists will be able to build entire microfactories that, powered by light, can make copies of themselves in mere minutes.

nologists aim to build materials and machinery by placing each atom and molecule precisely where they want it.

By such methods we could make truly identical parts and thus escape from the randomness that hinders conventionally made machines. Today, for example, when we try to etch very small circuits, the sizes of the wires vary so much that we cannot predict their electrical properties. If we can locate each atom exactly, however, the behavior of those wires would be indistinguishable. This capability would lead to new kinds of materials that current techniques could never make; we could endow them with enormous strength or novel quantum properties. These products in turn could lead to computers as small as synapses, having unparalleled speed and efficiency.

Once we can use these techniques to construct a general-purpose assembly machine that operates on atomic scales, further progress should be swift. If it took one week for such a machine to make a copy of itself, we could have a billion copies in less than a year. These devices would transform our world. For example, we could program them to fabricate efficient solar-energy collecting devices and attach these to nearby surfaces. Hence, the devices could power themselves. We would be able to grow fields of microfactories in much the same way that we now grow trees. In such a future we will have little trouble attaining wealth; our trouble will be in learning how to control it. In particular, we must always take care to maintain control over those things (such as ourselves) that might be able to reproduce themselves.

If we want to consider augmenting our brains, we might first ask how much a person knows today. Thomas K. Landauer of Bellcore reviewed many experiments in which people were asked to read text, look at pictures and listen to words, sentences, short passages of music and nonsense syllables. They were later tested to see how much they remembered. In none of these situations were people able to learn, and later remember for any extended period, more than about two bits per second. If one could maintain that rate for 12 hours every day for 100 years, the total would be about three billion bits—less than what we can currently store on a regular five-inch compact disc. In a decade or so that amount should fit on a single computer chip.

Although these experiments do not much resemble what we do in real life, we do not have any hard evidence that people can learn more quickly. Despite common reports about people with "photographic memories," no one seems to have mastered, word for word, the contents of as few as 100 books or of a single major encyclopedia. The complete works of Shakespeare come to about 130 million bits. Landauer's limit implies that a person would need at least four years to memorize them. We have no well-founded estimates of how much information we require to perform skills such as painting or skiing, but I do not see any reason why these activities should not be similarly limited.

The brain is believed to contain on the order of 100 trillion synapses, which should leave plenty of room for those few billion bits of reproducible memories. Someday, using nanotechnology, it should be feasible to build that much storage space into a package as small as a pea.

Once we know what we need to do, our nanotechnologies should enable us to construct replacement bodies and brains that will not be constrained to work at the crawling pace of "real time." The events in our computer chips already happen millions of times faster than those in brain cells. Hence, we could design our "mind-children" to think a million times faster than we do. To such a being, half a minute might seem as long as one of our years and each hour as long as an entire human lifetime.

But could such beings really exist? Many scholars from a variety of disciplines firmly maintain that machines will never have thoughts like ours, because no matter how we build them, they will always lack some vital ingredient. These thinkers refer to this missing essence by various names: sentience, consciousness, spirit or soul. Philosophers write entire books to prove that because of this deficiency, machines can never feel or understand the kinds of things that people do. Yet every proof in each of those books is flawed by assuming, in one way or another, what it purports to prove—the existence of some magical spark that has no detectable properties. I have no patience with such arguments. We should not be searching for any single missing part. Human thought has many ingredients, and every machine that we have ever built is missing dozens or hundreds of them! Compare what computers do today with what we call "thinking." Clearly, human thinking is far more flexible, resourceful and adaptable. When anything goes even slightly wrong within a present-day computer program, the machine will either come to a halt or generate worthless results. When a person thinks, things are constantly going wrong as well, yet such troubles rarely thwart us. Instead we simply try something else. We look at our problem differently and switch to another strategy. What empowers us to do this?

On my desk lies a textbook about the brain. Its index has approximately 6,000 lines that refer to hundreds of specialized structures. If you happen to injure some of these components, you could lose your ability to remember the names of animals. Another injury might leave you unable to make long-range plans. Another impairment could render you prone to suddenly utter dirty words because of damage to the machinery that normally censors that type of expression. We know from thousands of similar facts that the brain contains diverse machinery. Thus, your knowledge is represented in various forms that are stored in different regions of the brain, to be used by different processes. What are those representations like? We do not yet know.

But in the field of artificial intelligence, researchers have found several useful means to represent knowledge, each better suited to some purposes than to others. The most popular ones use collections of "if-then" rules. Other systems use structures called frames, which resemble forms that are to be filled out. Yet other programs use weblike networks or schemes that resemble trees or sequences of planlike scripts. Some systems store knowledge in languagelike sentences or in expressions of mathematical logic. A programmer starts any new job by trying to decide which representation will best accomplish the task at hand. Typically a computer program uses a single representation, which, should it fail, can cause the system to break down. This shortcoming justifies the common complaint that computers do not really "understand" what they are doing.

What does it mean to understand? Many philosophers have declared that understanding (or meaning or consciousness) must be a basic, elemental ability that only a living mind can possess. To me, this claim appears to be a symptom of "physics envy"—that is, they are jealous of how well physical science has explained so much in terms of so few principles. Physicists have done very well by rejecting all explanations that seem too complicated and then searching instead for simple ones. Still, this method does not work when we are addressing the full complexity of the brain. Here is an abridgment of what I said about the ability to understand in my book *The Society of Mind*:

> *If you understand something in only one way, then you do not really understand it at all. This is because if something goes wrong you get stuck with a thought that just sits in your mind with nowhere to go. The secret of what anything means to us depends on how we have connected it to all the other things we know. This is why, when someone learns "by rote," we say that they do not really understand. However, if you have several different representations, when one approach fails you can try another. Of course, making too many indiscriminate connections will turn a mind to mush. But well-connected representations let you turn ideas around in your mind, to envision things from many perspectives, until you find one that works for you. And that is what we mean by thinking!*

I think flexibility explains why, at the moment, thinking is easy for us and hard for computers. In *The Society of Mind,* I suggest that the brain rarely uses a single representation. Instead it always runs several scenarios in parallel so that multiple viewpoints are always available. Furthermore, each system is supervised by other, higher-level ones that keep track of their performance and reformulate problems when necessary. Because each part and process in the brain may have deficiencies, we should expect to find other parts that try to detect and correct such bugs.

In order to think effectively, you need multiple processes to help you describe, predict, explain, abstract and plan what your mind should do next. The reason we can think so well is not because we house mysterious sparklike talents and gifts but because we employ societies of agencies that work in concert to keep us from getting stuck. When we discover how these societies work, we can put them inside computers, too. Then if one procedure in a program gets stuck, another might suggest an alternative approach. If you saw a machine do things like that, you would certainly think it was conscious.

This chapter bears on our rights to have children, to change our genes and to die if we so wish. No popular ethical system yet, be it humanist or religion-based, has shown itself able to face the challenges that already confront us. How many people should occupy the earth? What sorts of people should they be? How should we share the available space? Clearly, we must change our ideas about making additional children. Individuals now are conceived by chance. Someday, instead, they could be "composed" in accord with considered desires and designs. Furthermore, when we build new brains, these need not start out the way ours do, with so little knowledge about the world. What kinds of things should our "mind-children" know? How many of them should we produce, and who should decide their attributes?

Traditional systems of ethical thought are focused mainly on individuals, as though they were the only entities of value. Obviously, we must also consider the rights and the roles of larger-scale beings—such as the superpersons we term cultures and the great, growing systems called sciences—that help us understand the world. How many such entities do we want? Which are the kinds that we most need? We ought to be wary of ones that get locked into forms that resist all further growth. Some future options have never been seen: imagine a scheme that could review both your mentality and mine and then compile a new, merged mind based on that shared experience.

Whatever the unknown future may bring, we are already changing the rules that made us. Most of us will fear change, but others will surely want to escape from our present limitations. When I decided to write this chapter, I tried these ideas out on several groups. I was amazed to find that at least three quarters of the individuals with whom I spoke seemed to feel our life spans were already too long. "Why would anyone want to live for 500 years? Wouldn't it be boring? What if you outlived all your friends? What would you do with all that time?" they asked. It seemed as though they secretly feared that they did not deserve to live so long. I find it rather worrisome that so many people are resigned to die. Might not such people, who feel that they do not have much to lose, be dangerous?

My scientist friends showed few such concerns. "There are countless things that I want to find out and so many problems I want to solve that I could use many centuries," they said. Certainly immortality would seem unattractive if it meant endless infirmity, debility and dependency on others, but we are assuming a state of perfect health. Some people expressed a sounder concern—that the old ones must die because young ones are needed to weed out their worn-out ideas. Yet if it is true, as I fear, that we are approaching our intellectual limits, then that response is not a good answer. We would still be cut off from the larger ideas in those oceans of wisdom beyond our grasp.

Will robots inherit the earth? Yes, but they will be our children. We owe our minds to the deaths and lives of all the creatures that were ever engaged in the struggle called evolution. Our job is to see that all this work shall not end up in meaningless waste.

Sustaining Life on the Earth

Hope for an environmentally sustainable future lies in evolving institutions, technology and global concern.

• • •

Robert W. Kates

Can life be sustained on the earth? If life is simply organic matter capable of reproducing itself, then the answer is almost assuredly "yes." Through the ages, life on the earth has survived repeated catastrophes, including atmospheric change, the submergence and reemergence of continents, and collisions with asteroids. Life will almost surely go on at least until the final "dimming of the light" of a cooling sun. But if life on the earth is life as we know it, the mix of living things that fill the places we are familiar with, then the answer is almost assuredly "no." For human-induced modifications to the environment, including to the global biogeochemical and hydrologic cycles, rival nature's changes to the earth. Most of the transformations of the past 10,000 years have occurred in our lifetimes, as humans continue to alter their environment in increasingly diverse ways.

If by life we mean us, our species and the life that supports us, then the answer is "perhaps." For humans, life has never really been simply a progression onward and upward from the cave. Our numbers have grown by fits and starts, our civilizations have declined and fallen, and even our physique has fluctuated over time. But since the middle of the last century our population has quadrupled, and projections from the United Nations and the World Bank suggest that it will at least double again by the middle of the next century. Economic activity amplified by technology has already transformed the earth.

What will be the impact of such numbers of humans, their rapidly changing patterns of habitation and their growing production and consumption, on the natural systems that support life? If we can manage the transition to a warmer, more crowded, more connected but more diverse world, there may be promise of an environmentally sustainable future (see Figure 10.1).

A recurring vision of the growth of world population is the exponential curve, which Thomas Robert Malthus proposed would eventually plunge when some maximum is reached. But this image, reminiscent of an accelerating rocket climbing out of sight toward sudden disaster, is misleading. Edward S. Deevey, Jr., offered a different view 34 years ago [see "The Human Population," SCIENTIFIC AMERICAN, September 1960]. He estimated the size of human population back as far as the origin of our

Figure 10.1 POPULATION MEETS NATURE in Ålesund, Norway, an island city in the Sunnmøre district that serves as a center for trading and fishing. The inhabitants of Ålesund also work in engineering firms and on some small farms (*see island in background at right*). In this northern temperate region, use of renewable resources and careful planning of manufacturing, business, agriculture and housing create a human community sustainable within the environment.

species, plotting it on a logarithmic scale. Deevey's sensitive, extended analysis revealed three surges in the number of people (see Figure 10.2).

Each surge coincided with a remarkable technological revolution: the emergence of toolmaking, the spread of agriculture and the rise of industry. Each transformed the meaning of resources and increased the carrying capacity of the earth. Each made possible a period of exponential growth followed by a period of approximate stability. The toolmaking, or cultural revolution, which began around one million years ago, saw human numbers rise to five million. Over the next 8,000 years, as humans domesticated plants and animals and invented agriculture and animal husbandry, the population grew 100-fold, to about 500 million. Now in this, the third population surge, we already number 5.6 billion—at best the midpoint on a projection that shows a doubling or even a tripling before growth levels off again—only 300 years after the scientific-industrial revolution began.

Even the global trend masks the existence of a deeper level of complexity. From its probable start in Africa, human life has steadily spread to every corner of the globe, including Antarctica, where research bases have altered the barren landscape.

But although the potential for humans to survive and even flourish in the most inhospitable of places has been realized, the history of life in certain ancient areas has been one of notable fluctuation.

My colleagues Thomas R. Gottschang of the College of the Holy Cross, Douglas L. Johnson and Billie L. Turner II of Clark University and Thomas M. Whitmore of the University of North Carolina at Chapel Hill and I have studied the phenomenon. To do so, we tried to reconstruct long, continuous series of human habitation for those places where we could correlate archaeological and historical accounts. Our original goal was to extend the record of habitation in order to relate fluctuations in natural processes such as climatic variation or soil formation to more rapidly changing patterns of human activity. Combining our data, we were able to reconstruct a long-term population sequence for four ancient regions (see Figure 10.3): the Nile Valley (6,000 years), the Tigris-Euphrates lowlands of Iraq (6,000 years), the basin of Mexico (3,000 years), and the central Maya lowlands of Mexico and Guatemala (2,200 years).

These reconstructed population series all show periodic fluctuations in growth and decline; in none does population grow without interruption.

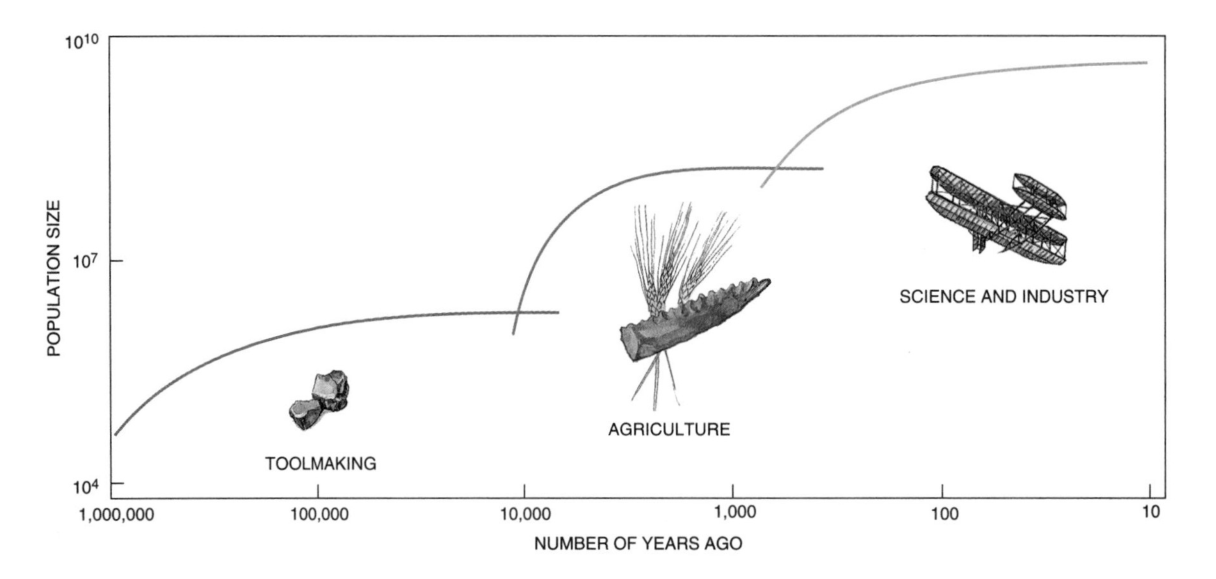

Figure 10.2 HUMAN POPULATION has grown dramatically over the past one million years. It has done so in three stages, each followed by a plateau. The first major growth, from 150,000 to five million, coincided with the development of toolmaking. The second surge, from five million to 500 million, was associated with the advent of agriculture. The third, from 500 million to 5.6 billion, is a consequence of the rise of industrial civilization. Each technological revolution—toolmaking, agriculture and manufacturing—has enabled humans to lessen their direct dependence on natural systems.

In all except the Maya case (the shortest record), there are 2.5 waves evident in which population at least doubled over the previous base and then fell by at least half with respect to that high point. The rates of growth and decline are modest in the early waves and more drastic in the later ones. The collapses of civilization, though surely catastrophic to the inhabitants, are not sudden. The second wave of declines, averaged for the four regions, lasts 500 years even though it includes one of the most precipitous extinctions in human history: the 16th-century epidemics among the native peoples of the New World.

Fluctuations in the well-being of entire civilizations are mirrored in the well-being of individuals. Again, seeking the long view of human life, my colleagues at Brown University—Robert S. Chen, William C. Crossgrove, Jeanne X. Kasperson, Robley Matthews, Ellen Messer, Sara R. Millman and Lucile F. Newman—and I considered human height. We assembled estimates, made by others, usually from the skeletons of adult males. We also considered studies of the measured heights of people from institutionalized populations. It is widely accepted that height, standardized by age and averaged over a population, reflects the state of nutrition and illness. In this way, times of hunger and ill health can be distinguished from those of plenty and wellness. Our analysis shows that throughout history, height, and presumably well-being, has fluctuated (see boxed figure "Height and Technological Change"). To take one example, an adult male in Roman Britain was as tall as or taller than his counterparts in this century, but his Victorian descendants were shorter. Thus, improvement in diet, health and sustenance has traced a halting, sometimes retrograde course.

These long waves of growth and decline in certain areas (which we have called millennial-long waves) raise questions about human life on the earth. Previously, the fates of particular places apparently averaged out—some developed, and others declined, with the overall balance one of punctuated growth. Has the scientific-industrial revolution, and the global economy to which it gave rise (complete with a global famine response system), exempted us from Malthusian-like collapses of the past? Or can particular regions, perhaps even regions that are world leaders, collapse in modern times?

Modern civilization has profoundly altered the environment. Concern about such effects has a history that extends back at least a century and a half [see "Origins of Western Environmentalism," by Richard Grove; SCIENTIFIC AMERICAN, July 1992]. As early as 1864, George Perkins Marsh published a benchmark assessment, *Man and Nature; or, Physical Geography as Modified by Human Action.* A subsequent account, entitled *Man's Role in Changing the Face of the Earth,* appeared in 1956. The most recent study, *The Earth as Transformed by Human Action,* was published in 1990.

An international collaborative effort, the Earth Transformed Project was seven years in the planning and execution. It brought together leading scientists from 16 countries to document global and regional change over the past 300 years. We were able to reconstruct human-induced change in 13 worldwide dimensions of chemical flow, land cover and biotic diversity: terrestrial vertebrate diversity, deforested area, soil area loss, sulfur releases, lead releases, carbon tetrachloride releases, marine mammal populations, water withdrawals, floral diversity, carbon releases, nitrogen releases, phosphorus releases and sediment flows.

The investigators took stock of the extent of human impact, emphasizing in particular the past 300 years. To place current changes in long-term perspective, we estimated human influence on the earth over the past 10,000 years, since the dawn of agriculture. In that time, humans have deforested a net area the size of the continental U.S., mostly using it for cropland. Water, in an amount greater than the contents of Lake Huron, is diverted every year from the hydrosphere for human use. Half the ecosystems of the ice-free lands of the earth have been modified, managed or utilized by people. The flows of materials and energy that are removed from their natural settings or synthesized now rival the flows of such materials within nature itself.

Most of this change has been extremely recent, considering that in seven of the 13 dimensions, half of all the change during the past 10,000 years happened within our lifetimes. To these rapid global environmental changes, it has now become fashionable to link threats emanating from political upheaval. Wars, especially in developing countries, are frequently attributed to famine, environmental disasters or scarcity of natural resources.

Here, where I live and write, on the coast of Maine, far from these disasters, I ask myself what might occur in the coming century. I have very good reasons to do so: six grandchildren who will be in their sixties and seventies by the year 2050. As I struggle to imagine their world, the ever present ocean suggests a metaphor of change that comes as currents, tides and surges. The currents are the long-term trends, the tides are the cyclical swings, and the surges, undertows and riptides are the surprises.

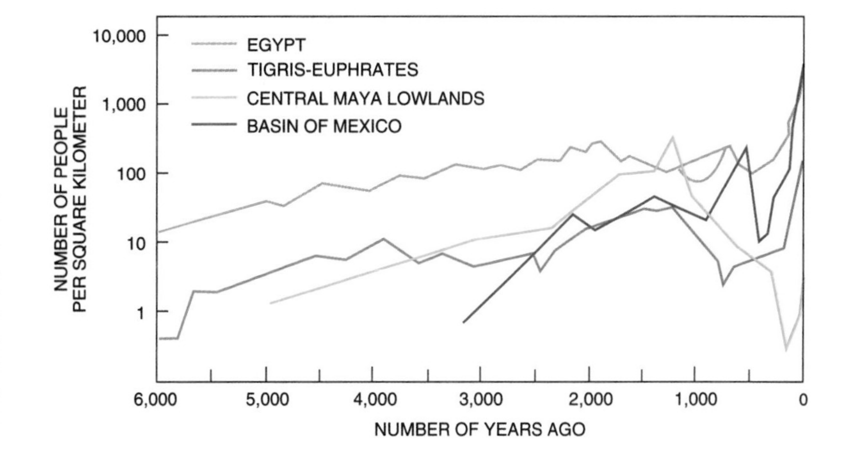

Figure 10.3 ADVANCE AND RETREAT of population density during historical times in four regions indicate that human numbers can fluctuate significantly. Civilizations in the Maya lowlands, the Tigris-Euphrates basin, the basin of Mexico and Egypt show periods of growth and decline. Is modern industrial society immune to this pattern?

Height and Technological Change

Average height, a standard measure of general well-being, has fluctuated over time. Hunter-gatherers in the eastern Mediterranean, who benefited from a diet full of calories and protein, reached a height of five feet, 10 inches. Early agriculturists from that area reached only five feet, three inches. They lived on a heavy cereal diet and suffered physical wear and tear from the difficult work of farming. Late agriculturists in Europe averaged five feet, nine inches. They presumably benefited from improvements in agricultural and other technologies. Stature fell again at the beginning of the European industrial period, when men reached five feet, seven inches on average. Modern U.S. men are slightly taller.

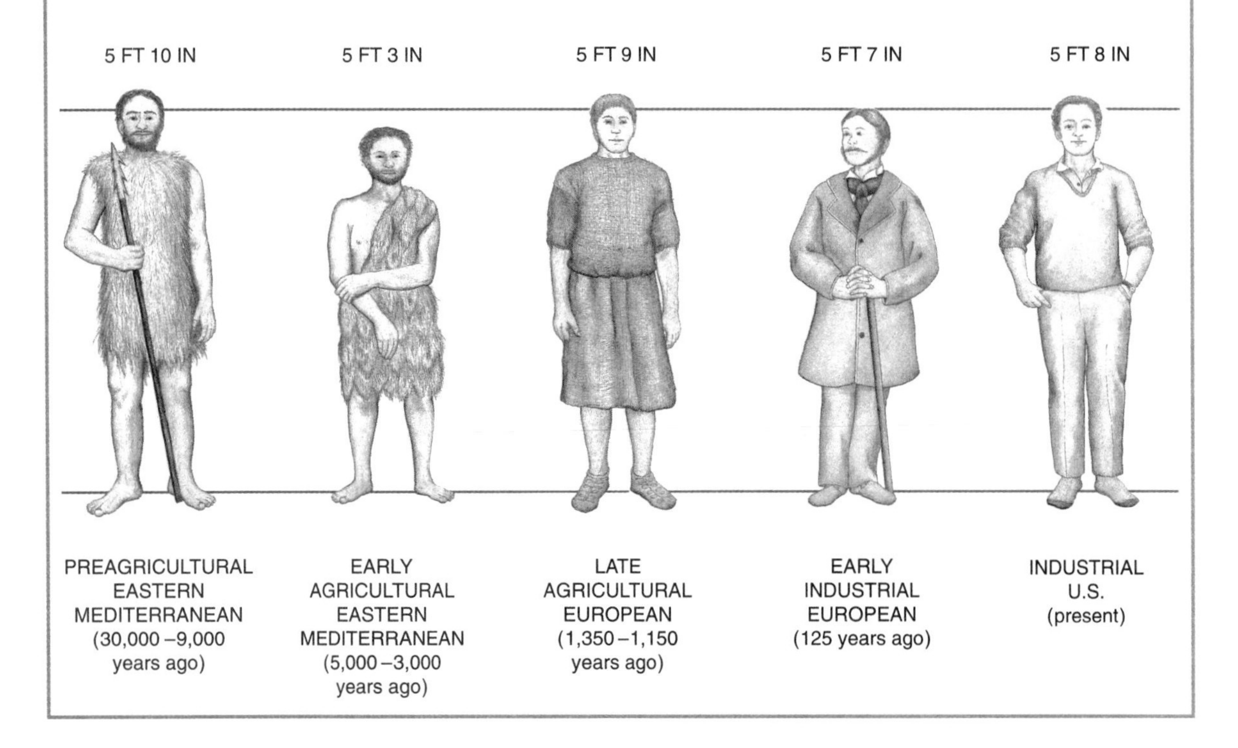

In particular, I believe the world of the next century will be warmer and more crowded, more connected but more diverse. Environmental change, population growth and increasing connectedness and diversity are powerful trends as deep-running as the ocean currents, seemingly set in place with little possibility of reversal, though clearly subject to slower or more rapid movement. Unless there is some flaw in present-day scientific understanding, we are already deeply committed to a warmer earth. Our world has been made more connected by a global economy and the widespread availability of rapid communication and transportation technology. This increasing connectedness will not necessarily homogenize people beyond their common affection for Coca-Cola, but it may, paradoxically, increase the diversity of both individuals and things. Goods, information and people are generally drawn to places of wealth or opportunity, which can make such areas more diverse. And strong countercurrents that emphasize ethnic, national and religious distinctiveness may create eddies and whirlpools where differing currents can mix and clash.

At the opposite extreme from the currents of certainty are the undertows, riptides and storm surges that batter our conventional expectations, leaving us only with the wisdom to expect surprises. National boundaries that had seemed immutable for decades have been swept away in a matter of months. Reaction to specific crises can deepen into new norms of human behavior and interaction. The spread of diseases such as AIDS can eat away at the foundations of society, increasing the potential for unforeseen disaster.

In contrast to the long-term trends and surprises, there are the short-term cycles or tides that are superimposed on the great underlying currents. As illustrations of the very short term, consider the oscillations of the business cycle or the so-called El Niño phenomenon that affects the Pacific Ocean and environs at irregular intervals of years. There are also decades-long fluctuations: in democracies, for instance, swings to the left or right of the political spectrum recur, as do periods of economic expansion and contraction.

Among these tides and storms, humans question their chances for long-term survival. Can our population continue to double and redouble within our children's and their children's lifetimes? Will there be food enough to feed the many, material sufficient for their needs and desires, and energy available to move and transform materials? Will the side effects of creating and using energy and of making and shaping materials undermine human health and destroy the ecological systems on which our species ultimately depends?

Such questions were powerfully posed 200 years ago by Malthus in his *Essay on the Principle of Population* (1798). They may be older yet: Tertullian wondered 1,800 years ago whether "pestilence, and famine, and wars, and earthquakes have to be regarded as a remedy for nations, as a means for pruning the luxuriance of the human race." It is not unexpected that Malthus, who was born in 1766 and died in 1834, worried about the adequacy of the resource base to feed England, because he lived in the midst of a population explosion. We now know that in the decade of his birth, England and Wales grew by 7 percent; in the decade of the first edition of his essay, by 11 percent. By the time of the fifth edition, in 1817, decadal growth had peaked at 18 percent.

Nor is it surprising that these concerns reemerge in the post–World War II world. The population explosion of the developing world was recognized in the late 1940s and the early 1950s. Indeed, the tides of scientific and public concern over population, food, materials, energy and pollution have surged, ebbed and surged again during the past 45 years, emerging most recently on the eve of the new millennium.

Among the threats that are most likely to occur, cause the most harm or affect the most people, I can identify three areas of concern. The first is the introduction of pollutants: acid rain in the atmosphere, heavy metals in the soils and chemicals in the groundwater. Humans also face the global atmospheric dangers of nuclear fallout, stratospheric ozone depletion and climatic warming from greenhouse gases. Finally, a massive assault on the biota has resulted from deforestation in the tropical and mountain lands, desertification in the drylands, and species extinction, particularly in the tropics.

A surge in production and consumption of material goods accompanies the rise in our numbers. In 1989 the International Institute of Applied Systems Analysis examined "current trends" or "business as usual" projections for a doubling of population. Its analysis assumed that varied and nutritious diets, industrial products and regular jobs are to be within reach of most of the 10 billion people. Thus, a doubling of the population will probably require a fourfold increase in agricultural production, a sixfold rise in energy use and an eightfold increase in the value of the global economy.

Many experts find this 2–4–6–8 scenario unbelievable and certainly unsustainable. Such increases, they think, could not be accommodated by present technology and practice in an environment that has already seen substantial transformation of its atmosphere, soils, groundwater and biota. Indeed, for many of today's Jeremiahs—Lester R. Brown, Paul R. and Anne H. Ehrlich, Donella H. Meadows, Dennis L. Meadows and Jørgen Randers—a world of more than five billion people is already overpopulated because virtually every nation is depleting its resources or degrading its environment. Other economists and technologists disagree [see "Can the Growing Human Population Feed Itself?" by John Bongaarts; SCIENTIFIC AMERICAN, March 1994]. They believe the invisible hand of rising prices will curb consumption and encourage conservation and invention.

They are confident that human creativity can overcome all limits.

But most of us, on reflection, recognize the unique situation that we face. In an extraordinarily short period—a matter of decades—society will need to feed, house, nurture, educate and employ at least as many more people as already live on the earth. If in such a warmer, more crowded world environmental catastrophe is to be avoided, it can be done only by maintaining severe inequities in human welfare or by adopting very different trajectories for technology and development.

How likely are we to have such different trajectories for technology and development? I draw cautious encouragement from two sets of trends that I perceive but do not fully understand. The first set relates to changes already apparent in the currents carrying us into the future. The second set relates to human adaptability in the form of the emergence of new institutions, technologies and, probably most important, ideas.

To illustrate some favorable changes in the currents, consider the *IPAT* equation. Initially formulated by Paul Ehrlich of Stanford University and John P. Holdren of the University of California at Berkeley, it is now widely used as a simplified statement of the driving forces of the human-induced detrimental impacts on the environment. The impacts term (I) is a function of population (P), the level of affluence (A) and the technology (T) available. Thus, the formulation captures the widespread agreement that to the extent that the environment is endangered, it is so not just because of the enormous growth of population (a common view in industrial countries) or just because of the rapacious and still growing use of energy and materials by affluent countries (a common view in poor countries). Instead both are significant contributing reasons. Estimates of the sources of greenhouse gases, for example, presume that most of these compounds originate in rich countries, but developing nations will contribute almost as much or more in 20 to 30 years if present trends persist. The technology term also captures the potential of science, technology and society to alter the impacts of any given level of population and affluence.

The growth of population and affluence and the spread of technologies are large-scale currents propelling us toward the warmer, more crowded, more connected but more diverse world. Countercurrents are already at work for each of the *IPAT* variables. Growth is slowing, and limits are in sight. Consider population and return to Deevey's vision of populations in flux. We are now in the last phase of the third major population surge, the completion of a demographic transition from a world with high rates of births and deaths to one with low rates. It took 150 years to complete this transition in England, but the transition in developing nations is occurring much more rapidly than expected.

Birth rates have fallen considerably from their post–World War II peak of five births per woman. The shift to 2.1 births per woman, required for zero-population growth, is just over halfway complete: the current birth rate is 3.2. The transition to low death rates is more advanced. In developing countries after World War II, the life expectancy at birth was 40 years. Now it has increased to 65 years, two thirds of the way to a likely average of 75 years, if we use developed countries as a model. The slowing of the rate of population growth everywhere, even very modestly in Africa, is a source of encouragement for sustaining life on the earth.

The affluence term may also be self-limiting: "Richer is cleaner." The Earth Transformed Project notes that the rates of increase for five of the 13 transformations studied have now turned downward: they are vertebrate and marine mammalian extinctions, as well as the release of lead, sulfur and carbon tetrachloride. All these have been the object of strenuous regulatory attention from the wealthier countries. A 1992 World Bank report argues that environmental problems shift with affluence. The poorest countries concentrate on primary needs for housing and sanitation, whereas in middle-income developing countries, efforts have shifted to grappling with air and water pollution. In the richest of countries, the focus has shifted from addressing localized problems to dealing with global environmental problems.

Economic and technological forces encourage reductions in the use of materials and energy in manufacturing: "Doing more with less." Since the mid-19th century the amount of carbon used per unit of production has been decreasing yearly by 1.3 percent through a combination of using less carbon-rich fuels (0.3 percent) to produce energy and using less energy overall per unit of production (1 percent). Nevertheless, these improvements in energy use have not been sufficient to offset the annual growth of the economy (3 percent). This has led to a global rise in carbon dioxide emissions of 1.7 percent every year.

A similar but more complicated trend toward dematerialization involves fewer materials per unit of production. We are using less steel and cement but more aluminum and chemicals (although use of the last two has peaked and is beginning to decrease). Despite the computer and television revolutions, the use of paper remains constant.

We should reconsider impacts as well. Scientists often do not sufficiently understand the effects of human-induced changes on the natural systems that support us to know how much or whether they are threatened or what replaces them when they are degraded. An apparent bias in research encourages the identification of harmful effects rather than the determination of negative feedback cycles that moderate the damage. For example, recent documentation shows that forest biomass in Europe is not only surviving but probably increasing, despite enormous burdens of pollutants and acid rain. That such a revitalization can happen, possibly through fertilization by the very same chemical pollutants that are causing the damage, is a caution. Nature may be more robust than popular rhetoric is willing to concede.

Optimists cite these countercurrents as good news. They argue that the trends, though insufficient to overcome the global growth in population or economy, are at least heading in the right direction. Pessimists either ignore the countercurrents or see them simply as too little and too late. It would help both sides to understand the many forces that are at work, invisible or otherwise. As yet, perception is quite dim. Consider the trend in population: What forces have lead to a decline in fertility?

A large amount of research has sought to estimate the dynamics and relative contributions of economic and social development and organized family-planning programs to the decline in births. Several studies, covering most developing countries, have found that increases in development are strongly associated with decline in birth rate, accounting for about two thirds of the drop. Additional research indicates that organized family-planning programs contribute another 15 to 20 percent to fertility reduction. Although only a few reports included culture and ethnicity, these factors also appear to be important. Socioeconomic development and substantial family-planning programs seem to be most effective in East and Southeast Asia or among those of Chinese extraction. Such programs are also effective when carried out on small, crowded islands or in city-states.

If, as the studies seem to show, "development is the best contraceptive," it is not clear which aspects of development are most influential. Analysts argue that as development proceeds, it lessens the need or desire for more children because more children survive, decreases the need for child labor and increases the need for educated children. Development also cuts the time available for childbearing and rearing and creates more opportunity for women to gain an education and find salaried work. Finally, it improves access to birth-control technology. Advocates for a particular policy usually single out one of these themes to justify their programs, but it is clear that better child survival, changing needs for labor, improved opportunities for women and access to birth control all occur together during the course of development.

Our understanding of the decline in fertility, as well as of the dynamics of pollution control and decarbonization of fossil fuels, remains opaque. For all these issues, there is no shortage of favored, oversimplified explanations to describe the countercurrents. And as with the fertility decline, a tension exists between separating the effects of such catchalls as "development" and "affluence" from the organized efforts of science or society.

The changes necessary to mitigate the extraordinary demands being placed on our life-support systems require a more fundamental understanding of our impacts on the earth. But the world does not and should not wait for such understanding. We know we must accelerate favorable trends and halt destructive ones. A species capable of questioning its own survival can also struggle to adapt to the warmer, more crowded and more connected world (see Figure 10.4). New institutions, new technologies and new ideas are already in place, defining a different trajectory of technology and development that a sustainable future might follow.

In a more connected but more diverse world, three sets of important transnational institutions are emerging. The best known are those created by governments, a set of international organizations, treaties and activities. Currently some 170 international treaties-in-force focus on the environment. New international institutions such as the United

Figure 10.4 GLOBAL VILLAGE includes American Samoa, where inhabitants watch television by the edge of the Pacific. Humankind needs to make a transition to a more crowded, warmer, more connected world while avoiding widespread catastrophe.

Nations Commission on Sustainable Development will oversee the accords of the 1992 Earth Summit in Rio de Janeiro. The Global Environmental Facility combines the talents and wealth of the World Bank and the U.N. Development (UNDP) and Environment (UNEP) Programs. Equally well known, but not usually for its environmental dimension, is the transnational corporation. Such corporations are responsible for many of the human-induced changes taking place around the globe; increasingly, however, they are also disseminators of common approaches, technical skills and standards for addressing environmental problems. Finally, least considered, but in many ways most important, is the veritable explosion of transnational nongovernmental and private voluntary environment and development organizations and their local counterparts in developing countries—an estimated 200,000 groups, increasingly linked together in international networks.

In a more crowded and more consuming world, one mode of coping is to use technology that requires modest amounts of such basic ingredients as energy, materials and information. As shown by long-term trends, there has already been a reduction of energy and materials required per unit of economic output, and the potential exists to accelerate such trends. Simple interventions include the recent competition to build a low-energy-consuming, non-ozone-depleting refrigerator, which will soon be on the market in the U.S. Another, related effort is the one to move immediately to next-generation refrigeration in India. In some eastern European countries, telephone companies are moving directly to wireless communications systems instead of rebuilding the fraying wire infrastructure.

The emergent field of study and action known as industrial ecology seeks to use the mechanisms of market competition and efficiency to minimize the amount of energy, materials and waste. Further into the future lies the substantial opportunity to increase human sustenance without increasing environmental burdens. The goal may be achieved through the science and engineering of biological processes, the development of new energy sources and transmission technologies, the creation of materials and, ultimately, the substitution of information for both energy and materials. Biotechnology promises crops that require less fertilizer and fewer pesticides. Researchers in the miniature world of nanotechnology and microelectronics hope to develop machines and processes that will require less bulk and thus less waste.

Potentially more significant than new institutions or technology are new ideas combined with an ever increasing concern for the environment. Sustaining human life on the earth requires at least three crucial sets of ideas: that cohabitation with the natural world is necessary; that there are limits to human activity; and that the benefits of human activity need to be more widely shared.

These ideas are spreading: last winter I heard a concert of 500 schoolchildren's voices, my grandchildren's included. Most striking to me was the relative absence of the patriotic songs of my childhood and their replacement by environmental hymns and anthems. I came away marveling at how 25 years of Earth Days have changed the formative ethos of young Americans. But this is Maine—what of the rest of the world?

A 1993 study undertaken by Riley E. Dunlap and the Gallup organization compared opinion on environmental issues in 12 industrial and 12 developing nations (including eastern Europe) and found surprisingly little difference in their attitudes (see boxed figure "Environmental Concern Is Global"). Even the attribution of the cause of the problems—

Environmental Concern Is Global

Despite the differences in population growth rates, available technology and level of affluence, people in industrial and developing nations share concern about the impact of human activities on the earth. A 1993 Gallup poll asked respondents in 24 countries, "How concerned are you personally about environmental problems?" Answers indicating "great concern" or a "fair amount of concern" generally dominated.

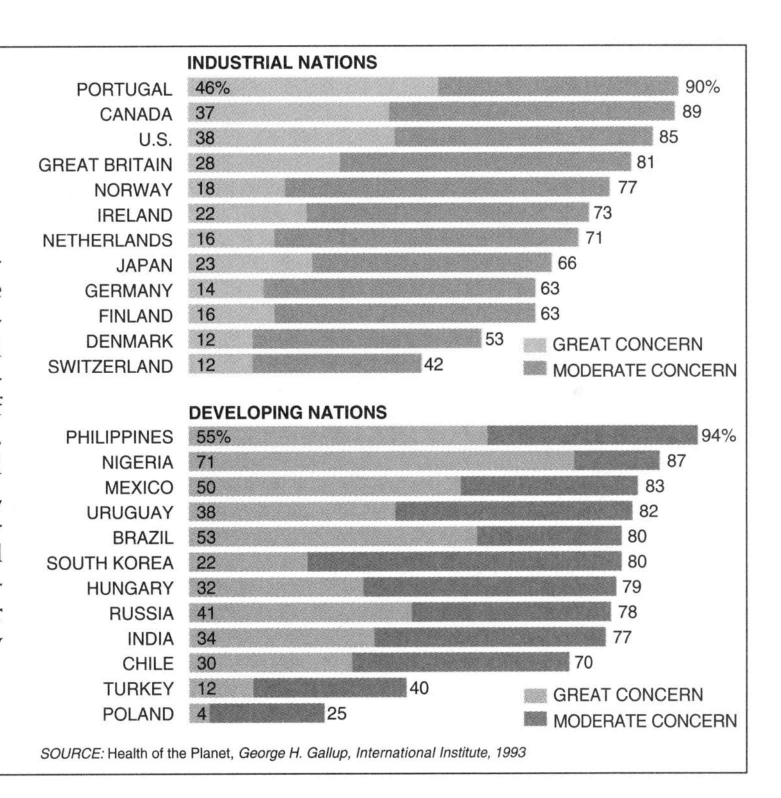

"overpopulation" and "consumption of the world's resources by industrial countries"—is seen as contributing equally by residents of both rich and poor countries. Along with this widespread evidence of environmental concern, more profound ideas are emerging. Witness the ongoing fundamental challenges to anthropocentrism and the more modest efforts to resolve the conflicting needs of ecosystems and economies or the conflicting claims of equity between species, places, peoples, livelihoods and generations.

Fifteen years ago Lionel Tiger of Rutgers University suggested that there was a "biology of hope," an evolutionary human tilt toward optimism that compensates in part for our ability to ask difficult questions such as "Can human life on the earth be sustained?" Although unpersuaded by his somewhat tenuous chains of argument, I share his inclination. Not because I have excessive confidence in the invisible hand of the marketplace, or of technological change, or even of James E. Lovelock's Gaia principle, in which life itself seems to create the conditions for its own survival. Nor is it just the wisdom and energy of my grandchildren and their enormous cohort of wise and energetic children around the world. Rather it is because hope is simply a necessity if we as a species, now conscious of the improbable and extraordinary journey taken by life in the universe, are to survive.

Epilogue

Descartes' Error and the Future of Human Life

• • •

Antonio R. Damasio

At the beginning of the 1950s, in an impassioned speech inspired by the threat of nuclear destruction, William Faulkner warned his fellow writers that they had "forgotten the problems of the human heart in conflict with itself." He asked them to leave no room in their workshops "for anything but the old verities and truths of the heart, the old universal truths lacking which any story is ephemeral and doomed—love and honor and pity and pride and compassion and sacrifice."

Although the towering nuclear threat of four decades ago has assumed a less dramatic posture, it is apparent to all but the most absent-minded optimists that other clear and present dangers confront us. The world population is still exploding; air, water and food are still being polluted; ethical and educational standards are still declining; violence and drug addiction are still rising. Many specific causes are at work behind all these developments, but through all of them runs the irrationality of human behavior, spreading like an epidemic, and no less threatening to our future than was the prospect of nuclear holocaust when Faulkner was moved to speak.

I have always taken his words to mean that the rationality required for humans to prevail and endure should be informed by the emotion and feeling that stem from the core of every one of us. This view strikes a sympathetic chord, because my research has persuaded me that emotion is integral to the process of reasoning. I even suspect that humanity is not suffering from a defect in logical competence but rather from a defect in the emotions that inform the deployment of logic.

What evidence can I produce to back these seemingly counterintuitive statements? The evidence comes from the study of previously rational individuals who, as a result of neurological damage in specific brain systems, lose their ability to make rational decisions along with their ability to process emotion normally. Their instruments of rationality can still be recruited; the knowledge of the world in which they must operate remains available; and their ability to tackle the logic of a problem remains intact. Yet many of their personal and social decisions are irrational, more often than not disadvantageous to the individual and to others. I have suggested that the delicate mechanism of reasoning is no longer affected by the weights that should have been imparted by emotion.

The patients so affected usually have damage to selected areas of the frontal, temporal and right parietal regions, but there are other conditions for which a neurological cause has not yet been identified, whose characteristics are similar in many respects. The sociopaths about whom we hear in the daily news are intelligent and logically competent individuals who nonetheless are deprived of normal emotional processing. Their irrational behavior is destructive to self and society.

Thus, absence of emotion appears to be at least as pernicious for rationality as excessive emotion. It certainly does not seem true that reason stands to gain from operating without the leverage of emotion. On the contrary, emotion probably assists reasoning, especially when it comes to personal and social matters, and eventually points us to the sector of the decision–making space that is most advantageous for us. In brief, I am not suggesting that emotions are a substitute for reason or that they decide for us. Nor am I denying that excessive emotion can breed irrationality. I am saying only that new neurological evidence suggests that no emotion at all is an even greater problem. Emotion may well be the support system without which the edifice of reason cannot function properly and may even collapse.

The idea that the bastion of logic should not be invaded by emotion and feeling is well established. You will find it in Plato as much as in Kant, but perhaps the idea would never have survived had it not been expressed as powerfully as it was by Descartes, who celebrated the separation of reason from emotion and severed reason from its biological foundation. Of course, the Cartesian split is not the cause of the contemporary pathologies of reason, but it should be blamed for the slowness with which the modern world has recognized their emotional root. When reason is conceptualized as free of biological antecedents, it is easier to overlook the role emotions play in its operation, easier not to notice that our purported rational decisions can be subtly manipulated by the emotions we want to keep at bay, easier not to worry about the possible negative consequences of the vicarious emotional experiences of violence as entertainment, easier to overlook the positive effect that well–tuned emotions can have in the management of human affairs.

It is not likely that reason begins with thought and language, in a rarefied cognitive domain, but rather that it originates from the biological regulation of a living organism bent on surviving. The brain core of complex organisms such as ours contains, in effect, a sophisticated apparatus for decisions that concern the maintenance of life processes. The responses of that apparatus include the regulation of the internal milieu, as well as drives, instincts and feelings. I suspect that rationality depends on the spirited passion for reason that animates such an apparatus.

It is intriguing to realize that Pascal prefigured this idea within the same 17th century that brought us Cartesian dualism, when he said, "It is on this knowledge of the heart and of the instincts that reason must establish itself and create the foundation for all its discourse." We are beginning to uncover the pertinent neurobiological facts behind Pascal's profound insight, and that may be none too soon. If the human species is to prevail, physical resources and social affairs must be wisely managed, and such wisdom will come most easily from the knowledgeable and thoughtful planning that characterizes the rational, self–knowing mind.

THE AUTHORS

STEVEN WEINBERG ("Life in the Universe") is a member of the physics and astronomy departments of the University of Texas at Austin. He was educated at Cornell University, the Niels Bohr Institute in Copenhagen and Princeton University and has received honorary degrees from a dozen other universities. His work has spanned a wide range of topics in elementary particle physics and cosmology, including the unification of the electromagnetic with the weak nuclear force, for which he shared the 1979 Nobel Prize for Physics. He is the recipient of the National Medal of Science (1991) and is a member of the National Academy of Sciences and Great Britain's Royal Society. In 1994 he served as president of the Philosophical Society of Texas.

P. JAMES E. PEEBLES, DAVID N. SCHRAMM, EDWIN L. TURNER and **RICHARD G. KRON** ("The Evolution of the Universe") have individually earned top honors for their work on the evolution of the universe. Peebles is professor of physics at Princeton University, where in 1958 he began an illustrious career in gravitational physics. Schramm is Louis Block Professor in the physical sciences department at the University of Chicago and the director of the Board of Physics and Astronomy of the National Research Council. Turner is associate chair of astrophysical sciences at Princeton and leads the council that oversees research at the Space Telescope Science Institute in Baltimore. Kron has served since 1978 on the faculty of the department of astronomy and astrophysics at Chicago and is a member of the experimental astrophysics group at the Fermi National Accelerator Laboratory.

ROBERT P. KIRSHNER ("The Earth's Elements") is chairman of the astronomy department at Harvard University. After receiving a Ph.D. in 1975 from the California Institute of Technology, he went to Kitt Peak National Observatory as a postdoctoral fellow. In 1976 he became an assistant professor at the University of Michigan and in 1985 moved to the Harvard-Smithsonian Center for Astrophysics. His work concentrates on supernovae and extragalactic astronomy. In 1992 he was elected a fellow of the American Academy of Arts and Sciences.

CLAUDE J. ALLÈGRE and **STEPHEN H. SCHNEIDER** ("The Evolution of the Earth") study various aspects of the earth's geologic history and its climate. Allègre is a professor at the University of Paris and directs the department of geochemistry at the Institut de Physique du Globe de Paris. He is a foreign member of the National Academy of Sciences. Schneider is a professor in the department of biological sciences at Stanford University and is a senior fellow at the university's Institute of International Studies. He is also a senior scientist at the National Center for Atmospheric Research in Boulder, Colorado.

LESLIE E. ORGEL ("The Origin of Life on the Earth") is senior fellow and research professor at the Salk Institute for Biological Studies in San Diego, which he joined in 1965. He obtained his Ph.D. in chemistry from the University of Oxford in 1951 and subsequently became a reader at the University of Cambridge, where he contributed to the development of ligand-field theory. He is also a fellow of the Royal Society and a member of the National Academy of Sciences.

STEPHEN JAY GOULD ("The Evolution of Life on the Earth") teaches biology, geology and the history of science at Harvard University, where he has been on the faculty since 1967. He received an A.B. from Antioch College and a Ph.D. in paleontology from Columbia University. Well-known for his popular scientific writings, in particular his monthly column in *Natural History* magazine, he is the author of 13 books.

CARL SAGAN ("The Search for Extraterrestrial Life") is a noted researcher, lecturer and author. He received his Ph.D. in astronomy and astrophysics from the University of Chicago in 1960. In 1968 he joined Cornell University, where he is David Duncan Professor of Astronomy and Space Science and director of the Laboratory for Planetary Studies. He has participated in numerous National Aeronautics and Space Administration planetary missions. His research interests embrace the origin of life, the physics and chemistry of planetary atmospheres and surfaces, and the search for extraterrestrial intelligence. His books include *Cosmos* and *Pale Blue Dot: A Vision of the Human Future in Space.*

WILLIAM H. CALVIN ("The Emergence of Intelligence") is a theoretical neurophysiologist at the University of Washington School of Medicine with a long association with neurosurgeons, zoologists and psychiatrists. He studied physics at Northwestern University, then neuroscience at the Massachusetts Institute of Technology and Harvard Medical School. He received his Ph.D. in physiology and biophysics from the University of Washington in 1966. He is also the author of several science books for the general public, including *The River That Flows Uphill, The Cerebral Symphony, The Ascent of Mind* and *How the Shaman Stole the Moon.*

MARVIN MINSKY ("Will Robots Inherit the Earth?") is Toshiba Professor of Media Arts and Sciences at the Massachusetts Institute of Technology. A pioneer in artificial intelligence and robotics, he began his career studying mathematics, physics, biology and psychology at Harvard and Princeton universities. In 1951 he designed and built with a colleague the first neural network learning machine. In the same decade he invented the confocal scanning microscope. He co-founded the Artificial Intelligence Laboratory at M.I.T. and does research at the Media Laboratory. He is also the author of numerous articles and books, including *The Society of Mind* (1987).

ROBERT W. KATES ("Sustaining Life on the Earth") is a geographer and independent scholar and University Professor (Emeritus) at Brown University. Between 1986 and 1992 he directed the Alan Shawn Feinstein World Hunger Program at Brown. He is also executive editor of *Environment* magazine, past president of the Association of American Geographers and co-chairman of the project Overcoming Hunger in the 1990s. He is the recipient of the National Medal of Science (1991) and a MacArthur Prize Fellowship (1981 to 1985) and a member of the National Academy of Sciences. His most recent works include co-authorships of *Population Growth and Agricultural Change in Africa* and a new edition of *The Environment as Hazard.*

ANTONIO R. DAMASIO ("Epilogue: Descartes' Error and the Future of Human Life") is M. W. Van Allen Professor and head of the neurology department at the University of Iowa College of Medicine. He is also an adjunct professor at the Salk Institute for Biological Studies in San Diego. His most recent work is *Descartes' Error: Emotion, Reason, and the Human Brain.*

BIBLIOGRAPHY

2. The Evolution of the Universe

Overbye, Dennis. 1991. *Lonely hearts of the cosmos: The scientific quest for the secret of the universe.* HarperCollins.

Riordan, Michael, and David N. Schramm. 1991. *The shadows of creation: Dark matter and the structure of the universe.* W. H. Freeman and Company.

Lemonick, Michael D. 1993. *The light at the edge of the universe: Astronomers on the front lines of the cosmological revolution.* Villard Books.

Peebles, P. J. E. 1993. *Principles of physical cosmology.* Princeton University Press.

3. The Earth's Elements

Ferris, Timothy. 1988. *Coming of age in the Milky Way.* William Morrow.

Murdin, Paul. 1990. *End in fire: The supernova in the Large Magellanic Cloud.* Cambridge University Press.

Kirshner, Robert P. 1992. Supernovae and stellar catastrophe. In *Understanding catastrophe,* ed., J. Bourriau. Cambridge University Press.

Bartusiak, Marcia. 1993. *Through a universe darkly: A cosmic tale of ancient ethers, dark matter, and the fate of the universe.* HarperCollins.

4. The Evolution of the Earth

Broecker, Wallace. 1990. *How to build a habitable planet.* Lamont-Doherty Geological Observatory Press.

Schneider, Stephen H., and Penelope J. Boston, eds. 1991. *Scientists on GAIA.* MIT Press.

Allègre, Claude J. 1992. *From stone to star: A view of modern geology.* Harvard University Press.

Kasting, James F. 1993. Earth's early atmosphere. *Science* 259 (February 12): 920–926.

5. The Origin of Life on the Earth

Miller, Stanley L., and Leslie E. Orgel. 1974. *The origins of life on the Earth.* Prentice-Hall.

Cairns-Smith, A. Graham. 1982. *Genetic takeover and the mineral origins of life.* Cambridge University Press.

Joyce, Gerald F. 1992. Directed molecular evolution. *Scientific American* 267 (December): 90–97.

Schopf, J. W. 1992. The oldest fossils and what they mean. In *Major events in the history of life,* ed., J. S. Schopf. Jones and Bartlett.

Gesteland, Raymond F., and John F. Atkins, eds. 1993. *The RNA world.* Cold Spring Harbor Laboratory Press.

6. The Evolution of Life on the Earth

Whittington, Henry B. 1985. *The Burgess shale.* Yale University Press.

Stanley, Steven M. 1987. *Extinction: A Scientific American book.* W. H. Freeman and Company.

Gould, Stephen Jay. 1989. *Wonderful life: The Burgess shale and the nature of history.* W. W. Norton.

Gould, Stephen Jay, ed. 1993. *The book of life.* W. W. Norton.

7. The Search for Extraterrestrial Life

Kieffer, H. H., B. M. Jakosky, C. Snyder and M. S. Matthews, eds. 1992. *Mars.* University of Arizona Press.

Sagan, Carl, W. Reid Thompson and Bishun N. Khare. 1992. Titan: A laboratory for prebiological organic chemistry. *Accounts of Chemical*

Research 25 (July): 286–292.

Horowitz, Paul, and Carl Sagan. 1993. Five years of Project META: An all-sky narrow-band radio search for extraterrestrial signals. *Astrophysical Journal* 415 (September 20): 218–233.

Sagan, Carl, et al. 1993. A search for life on Earth from the Galileo spacecraft. *Nature* 365 (October 21): 715–721.

8. The Emergence of Intelligence

Calvin, William H. 1991. *The ascent of mind: Ice age climate and the evolution of intelligence.* Bantam Books.

Savage-Rumbaugh, E. Sue, Jeannine Murphy, Rose A. Sevcik, Karen E. Brakke, Shelley L. Williams and Duane Rumbaugh. 1993. *Language comprehension in ape and child.* University of Chicago Press.

Gibson, Kathleen R., and Tim Ingold, eds. 1993. *Tools, language and cognition in human evolution.* Cambridge University Press.

Calvin, William H., and George A. Ojemann. *Conversations with Neil's brain: The neural nature of thought and language.* Addison-Wesley Publishing.

Pinker, Steven. 1994. *The language instinct.* William Morrow.

Khalfa, Jean, ed. 1994. *What is intelligence?* Cambridge University Press.

9. Will Robots Inherit the Earth?

Walford, Roy L. 1983. *Maximum life span.* W. W. Norton.

Minsky, Marvin. 1987. *The society of mind.* Simon and Schuster.

Moravec, Hans. 1988. *Mind children: The future of robot and human intelligence.* Harvard University Press.

Drexler, K. Eric. 1992. *Nanosystems.* John Wiley & Sons.

Minsky, Marvin, and Harry Harrison. 1992. *The Turing option.* Warner Books.

10. Sustaining Life on the Earth

Turner, B. L., II, William C. Clark, Robert W. Kates, John F. Richards, Jessica T. Mathews and William B. Meyer, eds. 1990. *The Earth as transformed by human action: Global and regional changes in the biosphere over the past 300 years.* Cambridge University Press.

Meadows, D. H., D. L. Meadows and J. Randers. 1992. *Beyond the limits: Confronting global collapse, envisioning a sustainable future.* Chelsea Green.

Ausubel, Jesse H., and H. Dale Langford, eds. 1994. *Technological trajectories and the human environment.* National Academy Press.

Sources of the Photographs

Alan Dressler, Carnegie Institute/National Aeronautics and Space Administration: NASA: Figure 2.1
NASA: Figure 2.3 *(left)*

J. Tesler, Arizona State University/NASA: Figures 3.1 and 3.5
Lawrence Livermore National Laboratory: Figure 3.2
NASA: "Supernova 1987A and the Age of the Universe"

Tim Fuller: Figure 4.4

Oscar Miller, Science Photo Library/Photo Researchers, Inc.: Figure 5.1
John Reader, Science Photo Library/Photo Researchers, Inc.: Figure 5.5 *(left)*
J. William Schopf: Figure 5.5 *(right)*

Mark McMenamin, Mount Holyoke College: Figure 6.1

Ken Biggs/Tony Stone: "What Is Life?" *(left)*
L'aura Colan/Photonica: "What Is Life?" *(center)*
Barry Parker/Bruce Coleman, Inc.: "What Is Life?" *(right)*
NASA: Figures 7.1, 7.3 and 7.4 *(top right)*
Paul Horowitz, Harvard University: "Does Intelligent Life Exist on Other Worlds?"
Mary Dale-Bannister, Washington University: Figure 7.2
Carl Sagan: Figure 7.4 *(left)*

Michael Nichols/Magnum Photos: Figures 8.2 and 8.4
Al Tielemans/Duomo: Figure 8.6

Kim Taylor/Bruce Coleman, Inc.: Figure 9.2
Sam Ogden: Figure 9.3
Electronic Design Center, Department of Electrical Engineering and Applied Physics, Case Western Reserve University: Figure 9.5

Koji Yamashita/Panoramic Images: Figure 10.1
Thomas Nebbia: Figure 10.4

INDEX

Page numbers in *italics* indicate illustrations.